Crash Course Astronomy

A Study Guide of Worksheets for Astronomy

By Roger Morante

Library of Congress Cataloging-in-Publication Date is available.

ISBN: 978-1-7359825-1-9
Writer: Roger Morante
Cover Design: Adobe stock image Purple Nebula #302158026 Purchased under Extended License
Cover Artist: Aaniyah Ahmed
Editor: Roger Morante
Copy Editor: Erica Brown
Back Cover Photo: Liesl Morante
Publisher Logo: Isabella Morante

To contact the publisher, send an email to the address below:
morante13@crashcourse.org.in

Additional copies may be purchased on Amazon.com or by contacting the author.
Printed in the United States of America
First printing April 2021

Table of Contents

Introduction to Astronomy: Crash Course Astronomy #1

1) Describe why people should make guesses about the **universe** to aid in the understanding of how **science** is a body of knowledge that may not be perfectly known.

_____
_____
_____
_____
_____
_____

2) Analyze the reasons why the study of **astronomy** is important to show not only how **science** is being constantly updated with new **theories** but also to help humans understand their place in the **universe**.

_____
_____
_____
_____
_____

3) Clarify why the study of **astronomy** is both the study of the *stuff* in the sky as well as the inclusion of a multitude of other fields in **science**.

_____
_____
_____
_____
_____

4) Connect the role that **astronomers** play in understanding the **universe**.

_____
_____
_____
_____
_____

5) Assess why the world needs **astronomers** and **engineers** as well as **software programmers**, **teachers**, **writers**, **video makers**, and **artists** to better understand the nature of the **universe**.

_____
_____
_____
_____
_____
_____

6) Draw conclusions as to how people in ancient times viewed the **stars** and the **sky**.

_____

_____

_____

_____

7) Explain **phenomena** in terms of concepts to understand how people in ancient times reacted when certain **stars** in the **sky** would appear when the days were longer, and certain **stars** would appear in the **sky** when the days were shorter.

_____

_____

_____

_____

_____

8) Differentiate between **astrology**, or the study of how the **positioning of the stars** and planets affect events on **Earth**, and **astronomy**, which studies everything outside of the **atmosphere** of the **Earth** to create a culture of the **stars**.

_____

_____

_____

_____

_____

9) Cite evidence to prove how **geocentrism** could make perfect sense in ancient times even though it is well known today that the **Earth** revolves around the **Sun** which in turn revolves around the **Milky Way Galaxy**.

_____

_____

_____

_____

_____

10) Analyze how the **polymath** Nicolaus Copernicus (1473-1543) came up with the idea that the **Sun** was the center of the solar system and not the **Earth** even though **geocentrism** had been the correct understanding of the **universe** for **millennia**.

_____

_____

_____

_____

_____

11) Briefly explain how **photography** and **calculus** combined to change our collective understanding of the **universe**.

_____

_____

_____

_____

Naked Eye Observations: Crash Course Anatomy #2

1) Connect how **naked eye** astronomy shaped the way humans thought about the **cosmos** thousands of years prior to the invention of the **telescope**.

_____

_____

_____

2) Analyze how **intrinsic physical brightness** and **distance** makes some **stars** seem very bright while others are not so bright when looking at the night sky.

   A) Factor 1 - **size**

   _____

   _____

   _____

   _____

   B) Factor 2 – **distance**

   _____

   _____

   _____

   _____

3) Investigate the importance around the creation of the first **star catalog** by the astronomer Hipparchus (190-120 BCE) who ranked stars by **luminosity** and **brightness**.

_____

_____

_____

_____

4) Explain why the star **Vega** appears to be blue while the star **Betelgeuse** appears to be red. *Include why **Arcturus** appears orange, **Capella** yellow, while fainter **stars** are white.*

_____

_____

_____

_____

5) Why do YOU think that the ancients would divide the stars into **constellations** and name them after familiar objects?

_____

_____

_____

_____

_____

6) Explain the problem with the **Big Dipper** as being inside of the **constellation** in the northern sky known as **Ursa Major**.

_____

_____

7) Clarify how stars in any **constellation** are given Greek letters in order of their **brightness**.

_____

_____

_____

8) Analyze the reasons why **light pollution** is a serious problem for **astronomers**. _Connect why **observatories** tend to be built in remote areas away from **light pollution**._

_____

_____

_____

_____

9) Synthesize solutions as to how people can cut back on **light pollution**. _Cite evidence of what has already been done to cut back on **light pollution**._

_____

_____

_____

10) Explain why **distortion** in the **atmosphere** of **Earth** causes **stars** to twinkle whereas the **planets** do NOT twinkle. _Include the **five planets** you can easily see in the **night sky**._

_____

_____

_____

_____

11) Briefly explain why **stars** move across the **sky** at night and connect why they appear to be invisible during the **day**. _Cite evidence to support your answer._

_____

_____

_____

_____

12) Analyze the reasons why the star **Polaris**, located in **Ursa Minor**, is called the **North Star**.

_____

_____

_____

_____

Cycles in the Sky: Crash Course Astronomy #3

1) Briefly explain why **stars** appear to move lower in the **horizon** after sunset and appear that way night after night changing their position in the **horizon** week after week.

   _____
   _____
   _____
   _____

2) Analyze the reasons why **stars** appear to **rise** and **set** at different times over the course of the year.

   _____
   _____
   _____
   _____
   _____
   _____

3) Develop a logical argument as to why the **stars** in the sky seem to follow an invisible line called the **ecliptic**.

   _____
   _____
   _____
   _____

4) Explain why the **constellations** in the sky appear to return to a certain **constellation** of the **zodiac** year after year.

   _____
   _____
   _____
   _____

5) Analyze how the **Earth** and its annual journey around the **Sun** creates the illusion of movement through the **zodiac** in the **night sky**.

   _____
   _____
   _____
   _____

6) Clarify why **planets** such as **Venus** and **Mercury** move so rapidly across the **night sky**.

   _____
   _____
   _____
   _____
   _____

7) Evaluate the **seasonal effect** the **tilting** of the **Earth** by 23.5 degrees on its **axis** has in relation to its **orbital plane**.

_____
_____
_____
_____
_____
_____
_____

8) Synthesize a scenario where the **Earth** does NOT have any **tilt** on its **axis**, and then hypothesize the effect **light** and **heat** from the **Sun** would have on **Earth**.

_____
_____
_____
_____
_____
_____
_____

9) Analyze the **26,000-year precession cycle** of the **Earth** to explain why **Polaris** will NOT be the **North Star** in **11,000 years** as that honor will be passed on to the star **Vega**.

_____
_____
_____
_____
_____
_____
_____

10) Clarify the reasons why the **astrological sign** of a person does NOT match where the **Sun** is in the sky due to the **precession** that has occurred over time. _Defend your answer with examples._

_____
_____
_____
_____
_____

11) Draw conclusions as to why **stars** were so important to **ancient humans**.

_____
_____
_____
_____
_____
_____

Moon Phases: Crash Course Astronomy #4

1) Briefly explain what causes the behavior of the **Moon** to change not only its **shape** but also its **location** in **space**.

_____

_____

_____

_____

2) Clarify why there are different **phases** of the **Moon** as viewed from **Earth**.

_____

_____

_____

_____

_____

3) Develop a logical argument as to why the **Moon** is always half-lit even though it has different **phases** as a result of its movement around the **terrestrial sky**.

_____

_____

_____

_____

4) Explain what happens when a **New Moon**, or the beginning of its **first cycle**, occurs on **Earth**.

_____

_____

_____

_____

5) Explain why we see the **terminator** around the **Moon** appear curved forming a thin illuminated **crescent Moon** from our **vantage point** on **Earth**.

_____

_____

_____

_____

_____

6) Analyze the reasons why the next **phase of the Moon** after the appearance of a thin **crescent Moon** is a **waxing crescent Moon**.

_____

_____

_____

_____

7) Explain why the **half-moon** is in its **first quarter phase** and we are looking straight down on its **terminator** after the first seven days of in its **29-day cycle**.

_____
_____
_____
_____
_____
_____

8) Differentiate between a **waxing gibbous Moon** and a **waning gibbous Moon**:
   a) **Waxing gibbous Moon** - lit side gaining – BEFORE **full Moon**

   _____
   _____
   _____
   _____
   _____

   b) **Waning gibbous Moon** – lit side shrinking – AFTER **full Moon**

   _____
   _____
   _____
   _____
   _____

9) Analyze why the **Moon** is half lit with its **terminator** appearing to split the face of the **Moon** into even halves during its **third quarter phase**.

_____
_____
_____
_____
_____

10) Briefly analyze why the **Moon** is considered to be a **waning crescent Moon** during its **third quarter phase**.

_____
_____
_____
_____
_____

11) Draw conclusions as to why the **phase** of the **Moon** has reached the point of being a **new Moon** again one month after a **new Moon**.

_____
_____
_____
_____
_____

Eclipses: Crash Course Astronomy #5

1) Connect how it is coincidental that the both the **Sun** and the **Moon** appear to be the same size in the sky even though the **Sun** is about 400 times wider and 400 times farther away than the **Moon** from **Earth**.

_____
_____
_____
_____

2) Differentiate between a **solar eclipse** and a **lunar eclipse**:
   a) **Solar eclipse** – localized on **Earth**

   _____
   _____
   _____

   b) **Lunar eclipse** - **Moon** in the shadow of **Earth**

   _____
   _____
   _____

3) Analyze the reasons why **Earth** does NOT have a **solar eclipse** every **new Moon** and a **lunar eclipse** every **full Moon**. *Include how the tilt of the **orbit** of the **Moon** around the **Earth** plays a role in this celestial dance.*

_____
_____
_____
_____

4) Explain the differences between how a **penumbra** and an **umbra** behave in casting shadows onto the surface of the **Earth** during an **eclipse**.
   a) **Penumbra**

   _____
   _____
   _____

   b) **Umbra**

   _____
   _____
   _____

5) Analyze what happens during a **total solar eclipse**.

_____
_____
_____
_____

6) Evaluate the reasons why the visible **corona** during a **total solar eclipse** causes the **atmosphere of the Sun** to seem even more magical than the day turning into night.

_____

_____

_____

7) Clarify why the term **Baily's beads** are used to describe the bright patches of **eclipsed Sun** along the edges of the **Moon** during an **eclipse**.

_____

_____

_____

8) Explain why a **total solar eclipse** creates a **diamond ring effect** around the **Moon**.

_____

_____

_____

9) Analyze how the **orbit** of the **Moon** around the **Earth** can affect its **penumbral shadow** during an **annual eclipse** and cause it to appear to be smaller than the **Sun** in the sky.

_____

_____

_____

_____

10) Evaluate the reasons why astronomers recommend extreme caution when viewing an **eclipse**. _Factor in **pupil dilation** which causes the **eyes** to let in an excess amount of light._

_____

_____

_____

_____

11) Describe the reasons why the **Moon** can turn a deep orange or blood red when it is inside the **penumbra** of the **Earth** during a **lunar eclipse**.

_____

_____

_____

_____

12) Explain the reasons why a **total lunar eclipse** can last nearly two hours whereas a **total solar eclipse** is over within minutes.

_____

_____

_____

_____

13) Clarify why it may be possible for **total eclipses** to stop happening in the future.

_____

_____

_____

Name_____
Period_____
Date_____

Telescopes: Crash Course Astronomy #6

1) Clarify the purpose of using a **telescope** in **astronomy**.

_____

_____

_____

_____

2) Compare how the bigger the **objective** (i.e. the higher quantity of **light** it collects) of a **telescope** to a bigger bucket that allows for the collection of higher quantities of water.

_____

_____

_____

_____

3) Explain how **light** in a **refractor telescope** becomes **concentrated** up to a tiny point that is then magnified by its **eyepiece** to focus the **light** of the **image** into your **eye**.

_____

_____

_____

_____

4) Use simple math to hypothesize what would happen if you doubled the **diameter** of the **collector** in a **telescope**. *Challenging – If you made the **light bucket** 10 times wider, how much **light** would be collected in the **telescope**.*

_____

_____

_____

_____

5) Explain the primary way in which **telescopes** work in collecting and redistributing **light**.

_____

_____

_____

_____

6) Describe how the **refraction** of **light** can make a spoon look like it is bending when it is observed inside a cup of **water**. *Cite evidence to back up your claims.*

_____

_____

_____

_____

_____

7) Connect how a **telescope** is able to make out **details** on planets such as **Jupiter** appear bigger than just observing **Jupiter** with your **naked eye**. *Include what happens to Venus, Saturn, Mars, and the Moon when observed through a refractor telescope.*

_____

_____

_____

_____

_____

8) Compare how **angular resolving power**, aka **resolution** of the **wave** nature of **light**, has the ability to separate two objects that are very close together (e.g. a distant car approaching YOU along a highway at night).

_____

_____

_____

_____

_____

9) Explain why **resolution** is more useful than **magnification** when looking through a **telescope**.

_____

_____

_____

_____

10) Analyze how scientist Isaac Newton (1642-1726) used **mirrors** to create a **reflector telescope**. *Clarify why he did this in order to solve the fuzzy problem in the optics when viewing distant objects and stars through a refractor telescope.*

_____

_____

_____

_____

11) Describe how objects can be seen in different types of **light** from **radio waves** to **gamma rays**, and then explain how this has aided humans in learning about the universe.

_____

_____

_____

_____

12) Evaluate how the era of **remote astronomy** has dramatically changed the field of **astronomy**.

_____

_____

_____

_____

_____

The Gravity of the Situation: Crash Course Astronomy #7

1) Explain the concept of how two objects can have the same **mass** but one can be much larger than the other. *Include how **density equals mass/volume** or **d=M/V***

_____

_____

_____

_____

2) Clarify how the **mass of an object** explains how much it resists having its **motion** changed.

_____

_____

_____

3) Point out the three things that explain the amount of **force** a person feels from the **gravity of an object**.

a) _____

b) _____

c) _____

4) Calculate how the force of **gravity** weakens with the **square of its distance**.

_____

_____

_____

5) Connect how **friction** and **air resistance** counteract **gravity**.

_____

_____

_____

6) Scientifically explain what would happen if a super person threw a rock hard enough to put it into **orbit** around the **Earth** by giving it **sideways motion**. *Include the role **gravity** would play on the rock while it was stuck in **orbit** around the **Earth**.*

_____

_____

_____

_____

7) Differentiate between a **circular orbit** and an **elliptical orbit** using the same idea as presented by the throwing of a rock around the **Earth** in question #6.

_____

_____

_____

_____

_____

8) Draw conclusions as to if it would be possible for a super person to throw a rock hard enough (**escape velocity**) for the rock to escape the **gravity** of **Earth**.

_____
_____
_____
_____

9) Differentiate as to why the **escape velocity** is greater for a person leaving **Jupiter** than for a person leaving **Earth**. _Include what the escape velocity is for the **Sun**._

_____
_____
_____
_____

10) Differentiate between a **parabola** and a **hyperbola** using the example of a super person throwing a rock from the surface of the **Earth**.

_____
_____
_____
_____

11) Explain why the force of **gravity** never quite drops to zero even though the force of **gravity** gets mathematically weaker with **distance**.

_____
_____
_____
_____

12) Clarify the reasons why **astronauts** on the **International Space Station** experience weightlessness even though **gravity** from **Earth** is still pulling on the **astronauts**. _Include why **NASA (National Aeronautics and Space Administration)** has given the forces acting upon the **International Space Station** and its inhabitants the name of **microgravity**._

_____
_____
_____
_____

13) ***BONUS QUESTION***
Calculate what YOUR **weight** would be on the **Moon, Mars, Jupiter**, or even the surface of the **Sun** using your current weight on **Earth** in both **pounds** and **kilograms**. _Explain why your **weight** would be not only different on these **celestial objects** but might also cause varying degrees of **balance** issue while moving on their surface._

_____
_____
_____
_____
_____

Tides: Crash Course Astronomy #8

1) Explain the reasons why **gravity** is measured from the **center of mass** of an **object**.

_____

_____

_____

_____

2) Analyze the multiple factors involved in order to determine how a **massive object** affects how strong its **tidal force** will be on an **affected object**.

a) **Gravity**

_____

_____

b) **Width**

_____

_____

c) **Distance**

_____

_____

3) Develop a logical argument which explains that when **gravity** is weaker on an **affected object** than so is the **tidal force**.

_____

_____

_____

_____

_____

4) Connect the reasons why the side of the **Earth** facing the **Moon** is pulled harder by **tidal forces** stretched by the **Moon** than by the opposite side of the **Earth**. *Hint – The diameter of the Earth is 12,742 kilometers or roughly 7917 ½ miles)*

_____

_____

_____

_____

_____

5) Clarify where the **tidal forces** are strongest on **Earth.** *Include how the relationship of the Earth to the position of the Moon plays a factor.*

_____

_____

_____

_____

_____

6) Explain the reasons why there are two **high tides** and two **low tides** that cause the **sea level** to rise and fall by a meter or two on a specific area of **Earth** each day.

_____
_____
_____
_____
_____
_____

7) Connect how the **Earth** can be affected by **tidal forces**.

_____
_____
_____
_____

8) Clarify how **gravity** and **inertia** act in opposition to the **oceans** of the **Earth.** *Include how both forces manifest themselves in* **tidal bulges** *on Earth caused by the* **gravitational force** *of the* **Moon***.*

_____
_____
_____
_____
_____

9) Explain how the **Moon** is not only getting farther away from **Earth** but also that the **Earth** is slowing in its **rotation**. *Include how* **friction** *created by the* **tidal forces** *is exchanged between the two* **celestial objects***.*

_____
_____
_____
_____
_____

10) Clarify how **tidal locking** prevents humans from viewing the **dark side of the Moon** while standing on **Earth**.

_____
_____
_____
_____
_____

11) Explain why the **Sun** does NOT affect **tidal forces** on **Earth** as strongly as the **Moon** yet can create different **tidal effects** on **Earth** during **spring tide** and **neap tide** events.

_____
_____
_____
_____
_____

Introduction to the Solar System: Crash Course Astronomy #9

1) Draw conclusions as to why people would choose to ignore ancient Greek astronomer and mathematician Aristarchus of Samos (310-230 BCE) when he presented the first known **heliocentric model** that placed the **Sun** rather than **Earth** at the center of the universe.

_____

_____

_____

_____

2) Clarify how the ancient historical argument made by Aristotle (384-322 BCE) of whether or not the **Earth** was at the center of the universe, aka a **geocentric universe**, and made during the **Classical period** in **Ancient Greece**, strongly influenced the **geocentric theories** of Ptolemy (100-170), which stuck around until the 16th century.

_____

_____

_____

_____

3) Analyze how the **heliocentric models** of Nicolaus Copernicus (1473-1543) and Johannes Kepler (1571-1630) paved the way for Isaac Newton (1642-1727) to apply both **physics** and **calculus** to explain how the **solar system** truly operates.

_____

_____

_____

_____

4) Differentiate between the **size** and **volume** of **Jupiter** as compared to the **Earth**.

_____

_____

_____

_____

5) Evaluate the **debate** over what constitutes a **planet** and what does NOT constitute a **planet**. *Defend your answer citing evidence to support YOUR ideas on the **planet debate**.*

_____

_____

_____

_____

6) Hypothesize as to why all the **planets** are swirling around the **Sun** as if it were on the plane of a **flat disk** rather than all around it like bees swarming around a hive.

_____
_____
_____
_____
_____

7) Differentiate between the basic **compositions** of the **inner rocky planets** (i.e. **Mercury, Venus, Earth,** and **Mars**) and the currently accepted compositions of the **outer gaseous planets** (i.e. **Jupiter, Saturn, Uranus,** and **Neptune**).

_____
_____
_____
_____
_____

8) Explain how the **solar system** formed out of a **cloud** which gave way to **planetesimals** and a **protostar** according to the current accepted theory of the 21st century.

_____
_____
_____
_____
_____
_____
_____

9) Analyze the reasons why the **inner planets** do NOT have a boatload of **hydrogen** and **helium** floating around in their **atmospheres** like that found on the **outer planets**.

_____
_____
_____
_____
_____

10) Cite evidence to back up the theory of the **formation of the solar system** as understood in the 21st century.

_____
_____
_____
_____
_____
_____
_____
_____
_____

The Sun: Crash Course Astronomy #10

1) Briefly describe the composition of the **Sun**.

_____
_____
_____
_____

2) Explain what happens to **hydrogen atoms** and their **electrons** due to the incredibly high **temperature** and crushing **pressure** inside the **core** of the **Sun**.

_____
_____
_____
_____
_____
_____
_____
_____

3) Analyze how **hydrogen** fuses into **helium** inside the **core** of the **Sun**. *Include how it creates **heat** that can be felt on **Earth**, as well as how **convection** in its interior causes it to shine so brightly in the sky that a person cannot look at the **Sun** with their naked eye.*

_____
_____
_____
_____
_____
_____
_____

4) Describe what is going on inside of the **photosphere** of the **Sun**.

_____
_____
_____
_____
_____

5) Analyze how the **corona** of the **Sun**, along with its **solar wind**, is similar to what could be considered the **atmosphere** of the **Sun**. *Include why its **corona** is so difficult to see.*

_____
_____
_____
_____
_____

6) Clarify how **light** is emitted from the **Sun** once it becomes a much **lower-energy photon** of **visible light**.

_____
_____
_____
_____

7) Clarify how long **light** from the **core** of the **Sun** takes to reach the **surface** of the **Earth**.

_____
_____
_____
_____

8) Analyze how a soup of rapidly moving charged **particles**, which begins its journey inside the **core** of the **Sun**, moves outwards to become **plasma** and in the process creates an intense **magnetic field** that also spawns **sunspots** on surface of the **Sun**.

_____
_____
_____
_____

9) Analyze how **sunspots** can heat up **plasma** to create **faculae** as well as being responsible for the increase in **energy output** emitted by the **Sun**.

_____
_____
_____
_____

10) Explain how **solar flares** occur when **magnetic fields** on the **surface of the Sun** snap.

_____
_____
_____
_____
_____

11) Differentiate between a **solar flare** and a **coronal mass ejection** (**CME**). _Include why there are **auroras** near the **poles** of the **Earth** when these events happen on the **Sun**._

_____
_____
_____
_____
_____

12) Analyze the reasons why there are **power outages** when **solar storms** hit the **Earth**.

_____
_____
_____
_____

Name_____
Period_____
Date_____

The Earth: Crash Course Astronomy #11

1) Explain how **liquid iron** in the **outer core** of the **Earth** is responsible for the **magnetic field** surrounding the planet.

_____
_____
_____
_____
_____

2) Describe why the **crust** of the **Earth** is NOT a solid piece. *Include why it is broken up into huge **plates** that move due to the flow of heated rock inside of the **mantle**.*

_____
_____
_____
_____

3) Describe how **convection** moves huge streams of warmer material such as rock through the **mantle**.

_____
_____
_____
_____

4) Clarify the reasons why **continents** on **Earth** are NOT where they were millions of years ago NOR where they will be millions of years from now.

_____
_____
_____
_____

5) Explain how the **formation** of the early **Earth** explains why its **core** is so hot. *Include an analysis from the leftover **heat** of **formation**, the squeeze of **gravity**, decaying **uranium**, and **friction**.*

_____
_____
_____

6) Evaluate how **convection** inside of the **liquid outer core** of the **Earth** conducts **electricity** that in turn generates a powerful **magnetic field** which surrounds the planet.

_____
_____
_____
_____

7) Analyze how the **magnetic field** of **Earth** deflects most of the **charged particles** (i.e. a **plasma** of **electrons, protons,** and **alpha particles**) blown out from **solar wind**.

_____
_____
_____
_____

8) Clarify what is known as the **Kármán line** or what is considered to be the barrier between the **atmosphere** of **Earth** and **space**.

_____
_____
_____
_____

9) Describe how **convection** in the air creates currents of rising air which can carry water with them to form **clouds**.

_____
_____
_____
_____

10) Analyze the reasons why the **ozone** is important in absorbing **solar ultraviolet light**.

_____
_____
_____
_____

11) Analyze how energized **oxygen** and **hydrogen** molecules about 150 km (kilometers) up in the **atmosphere** are responsible for the **aurora borealis effect**.

_____
_____
_____
_____

12) Connect how the **human body** has an **internal pressure** that balances out the ton of air that is pushing down upon the surface of the **Earth**.

_____
_____
_____
_____

13) Briefly analyze the reasons why the amount of **carbon dioxide** (CO_2) in the **atmosphere** is critical to the **greenhouse effect** as well as to the **biogenesis** of life on **Earth**.

_____
_____
_____
_____

The Moon: Crash Course Astronomy #12

1) Analyze the two factors of **Moon Illusion** to illustrate why the **Moon** is bigger on the **horizon** then it is directly overhead (i.e. **perception of objects, perception of the sky**).

_____
_____
_____
_____
_____
_____

2) Compare the internal structure of the **Moon** to the internal structure of the **Earth**.

_____
_____
_____
_____
_____
_____
_____
_____

3) Differentiate between the **Lunar Highland Regions** of the **Moon** and its **Lunar Maria Regions**.

_____
_____
_____
_____
_____
_____

4) Defend the **Giant Impact Hypothesis** concerning how the **Moon** formed and synthesize a few reasons that explains why the **composition** of the **crust** of the **Moon** has a lot of similarities to **Earth** but also a lot of differences.

_____
_____
_____
_____
_____
_____
_____
_____

5) Analyze the 21st century twist to the **Giant Impact Hypothesis** that explains how features on the **near side** and **far side** of the **Moon** formed after a Mars-sized object named **Theia** slammed into the early **Earth**.

_____
_____
_____
_____
_____
_____
_____

6) Hypothesize as to how the ancient Trojan planet **Theia** may have come from the outer solar system. *Connect how it could have brought **water** along with it.*

_____
_____
_____
_____
_____
_____

7) Explain what happened during the **Late Heavy Bombardment** (4.1-3.8 billion years ago) period on the **Moon**.

_____
_____
_____
_____
_____
_____
_____

8) Assess how **crater chains** formed from **asteroid** and **comet material**.

_____
_____
_____
_____
_____

9) Explain the reason why **rays**, such as those surrounding the crater **Tycho**, are common around big craters on the **Moon**.

_____
_____
_____
_____
_____

10) Design a scenario that includes what it would be like to live and survive inside a former **lava tube** on the **Moon**. *Include how mining **water ice** off the surface of the **Moon**, as well as the refilling of both **oxygen** and **nitrogen tanks**, would play a large role in YOUR lunar life.*

_____
_____
_____
_____
_____
_____
_____
_____
_____
_____
_____
_____
_____
_____
_____
_____
_____
_____
_____
_____
_____
_____
_____
_____
_____
_____
_____
_____
_____
_____
_____
_____
_____
_____

Mercury: Crash Course Astronomy #13

1) Explain how astronomer and Jesuit priest Giovanni Zupi (1589-1650) demonstrated that **Mercury** orbited the **Sun (heliocentrism)** and undergoes a complete cycle of phases just like the **Moon** in 1639.

_____

_____

_____

_____

2) Clarify how **gravity** from the **Sun** affects the **orbital velocity** of **Mercury**. *Mercury orbits the Sun in 88 days in contrast to the 365-day orbit of the Earth around the Sun.*

_____

_____

_____

_____

_____

3) Describe how the **orbit** of **Mercury** plays into how much **light** and **heat** it absorbs from the **Sun**.

_____

_____

_____

_____

_____

4) Explain why it proved difficult for **astronomers** to figure out how long a **day** lasted on **Mercury** since it is **tidally locked** to the **Sun**. *Include why astronomers assumed that Mercury initially spun like the Moon.*

_____

_____

_____

_____

_____

_____

5) Clarify the reasons why **tides** (created by the **Sun**) are far stronger on **Mercury** when the planet is at **perihelion** than when it is at **aphelion**.

_____

_____

_____

_____

_____

_____

6) Analyze how **tidal forces** from the **Sun** slowed the **rotation** of **Mercury** to the point that its spin slowed to 2/3 of its **orbital period** causing only one side of the planet to face the **Sun** for 88 days.

_____
_____
_____
_____

7) Clarify how it would be possible to watch the **Sun** rise, slow, stop, set again, and then rise again if you were standing on the right spot on the surface of **Mercury**.

_____
_____
_____
_____

8) Explain the reasons why **Mercury** is hard to observe from the **vantage point** of **Earth**.

_____
_____
_____
_____

9) Describe how **scarps**, aka compression folds such as that on the **thrust fault** of **Discovery Rupes**, formed on the planet **Mercury** likening its surface to a wrinkly fruit.

_____
_____
_____
_____

10) Clarify the names of a few **craters** on **Mercury** which just happen to be named after various artists.

_____
_____
_____
_____

11) Hypothesize the reasons why **Mercury** must have a large **iron core** as a result of its **formation** as a planet during the early days of the solar system.

_____
_____
_____
_____

12) Connect the reasons why **Mercury** has a **magnetic field**, albeit a weak one, even though it does not spin very fast.

_____
_____
_____
_____

Venus: Crash Course Astronomy #14

1) Clarify the reasons why the best time to look at **Venus** is after sunset or before sunrise as a result of the **transit of Venus** across the night sky.

_____
_____
_____
_____

2) Explain the reasons why scientists had to wait until **radar** was invented in the 20th century to calculate the exact distance to **Venus**.

_____
_____
_____
_____

3) Analyze the reasons why **Venus** is so **bright** that it can even be seen during the daytime.

_____
_____
_____
_____

4) Compare the similarities and differences between the planets **Venus** and **Earth**. *Explain why some people have called the planets **twins**.*

_____
_____
_____
_____
_____

5) Describe the **atmospheric conditions, pressures,** and **temperatures** on **Venus**.

_____
_____
_____
_____

6) Analyze the reasons why **conditions** on the **surface** of **Venus** are literally **Hell** and super-hot and have led to a **runaway greenhouse effect** on the planet.

_____
_____
_____
_____
_____
_____

7) Evaluate the problems the planet **Venus** has had during its lifetime because of the absence of a **magnetic field** and **lack of protection** from the **solar wind**.

_____
_____
_____
_____
_____
_____

8) Assess the reasons why **Venus** is way **hotter** than **Mercury** even though **Mercury** is closer to the **Sun**.

_____
_____
_____
_____

9) Briefly explain why it might be possible for **metal** to snow on **Venus**.

_____
_____
_____
_____

10) Hypothesize as to why **Venus** does NOT have a **magnetic field**.

_____
_____
_____
_____

11) Clarify the reasons why the **surface temperature** on **Venus** is about the same everywhere on the planet.

_____
_____
_____
_____

12) Differentiate between **plate tectonics** on the surface of **Venus** and **plate tectonics** on the **surface of Earth** due to scale of the **greenhouse effect** of **Venus**.

_____
_____
_____
_____

13) Analyze how **volcanic activity** on Venus could have played a role in spewing **lava** that may have covered the entire surface of **Venus** as recently as 500 million years ago. _Include how the entire planet of **Venus** could be dominated by a **super volcano**._

_____
_____
_____
_____

Mars: Crash Course Astronomy #15

1) Briefly explain why **Mars**, the fourth planet from the **Sun**, is the color red.

_____
_____
_____
_____
_____

2) Clarify the **type of planet** that **probes** sent to **Mars** by **scientists** in the 1960s and 1970s expected to see. *Include what they actually saw.*

_____
_____
_____
_____

3) Hypothesize how a few of the main features on **Mars** such as the mountain **Olympus Mons** in the **Tharsis Bulge** area formed on the **planet**.

_____
_____
_____
_____
_____

4) Draw conclusions as to how **Valles Marineris**, a canyon that is 4,000 kilometers long, 200 kilometers wide, and 7 kilometers deep, formed on the planet **Mars**.

_____
_____
_____
_____

5) Analyze how the **polar ice caps** on **Mars**, which are made out of mostly **water ice** (H_2O) along with a seasonal dusting of a **carbon dioxide** (CO_2) layer (**dry ice**), affect the **wind patterns** on **Mars**.

_____
_____
_____
_____

6) Describe the **atmosphere** and **pressure** on the **surface** of **Mars** in comparison to **Earth**.

_____
_____
_____
_____
_____

7) Analyze the **composition** of **sand** on **Mars**. *Include reasons why **Mars** has **sand dunes**.*

_____

_____

_____

_____

8) Explain how a mixture of **low pressure** and **carbon dioxide** causes massive **ice avalanches**, aka **fast running glacier surges**, during the **springtime** on the planet **Mars**.

_____

_____

_____

_____

9) Hypothesize as to how **Phobos** and **Deimos** formed into two small **moons** that **orbit** the planet **Mars**.

_____

_____

_____

_____

10) Develop a logical argument as to whether or not YOU think there is or has ever been **liquid water** on the **surface** of **Mars**. *Defend your answer.*

_____

_____

_____

_____

_____

11) Rationalize whether or not YOU think **Mars** has **life today** or has ever held **life** in the **past**. *Defend your answer.*

_____

_____

_____

_____

12) Connect how the presence of the molecule **methane (CH_4)** could have been emitted or exhaled by a **lifeform** on **Mars**.

_____

_____

_____

_____

13) Do YOU think that **people** will ever be able to **live** on and **colonize** the planet **Mars**? *Why or why not? Defend your answer.*

_____

_____

_____

_____

_____

Jupiter: Crash Course Astronomy #16

1) Clarify how long a **day** is on **Jupiter** on account of its **speed of rotation**, and then explain why that fact is so extraordinary.

_____
_____
_____
_____

2) Evaluate the statement by Galileo Galilei (1564-1642) that the **Galilean moons** of **Jupiter** (**Europa**, **Io**, **Ganymede**, and **Calisto**,) are "worlds in their own right."

_____
_____
_____
_____
_____
_____

3) Describe the formation of the **clouds** that people look at when they view **Jupiter** through a **telescope**.

_____
_____
_____
_____
_____

4) Differentiate between the **belts** and **zones** that circulate in opposite directions around **Jupiter**. _Include the motions, colors, formations, and compositions of them._

_____
_____
_____
_____
_____

5) Speculate as to the formation and sustainability of the **Jovian storm**, aka the **Giant Red Spot**, currently raging on **Jupiter** in the **21st century**.

_____
_____
_____
_____
_____
_____

6) Deconstruct the **composition** of the **Jovian atmosphere**.

_____
_____
_____
_____
_____

7) Analyze the **liquid hydrogen** electrical **surface composition** deep inside of **Jupiter**.

_____
_____
_____
_____
_____
_____

8) Analyze how the **Grand Tack** model explains how **Jupiter** formed in the early solar system.

_____
_____
_____
_____
_____

9) Explain why **Jupiter** emits more **heat** than it receives from the **Sun**.

_____
_____
_____
_____
_____
_____
_____

10) Connect how the **rapid rotation** of **Jupiter** along with its electrified **metallic hydrogen** liquid **surface** have led it to develop a very **strong magnetic field**.

_____
_____
_____
_____
_____
_____

11) Describe the effects that **comet Shoemaker-Levy 9** had on **Jupiter** when it crashed into the planet in 1994.

_____
_____
_____
_____

Jupiter's Moons: Crash Course Astronomy #17

1) Clarify why it was a pretty big deal when astronomer Galileo Galilei (1564-1642) observed that **Jupiter** had four moons (aka the Galilean Moons of **Io**, **Europa**, **Ganymede**, and **Callisto**) in 1610.

_____
_____
_____
_____
_____

2) Draw conclusions as to how the composition of **Ganymede** is likely responsible for its **magnetic field**. *Speculate as to what this might mean for potential life on that **moon**.*

_____
_____
_____
_____

3) Explain how scientists know that there is **subsurface water** on **Ganymede** even though no scientist in the first quarter of the 21st century has ever been to that **moon**.

_____
_____
_____
_____

4) Connect how the absence of **volcanoes** and **tectonic activity** on the heavily cratered surface of **Callisto** is a good indicator that it probably does NOT have a **metallic core** even though it has a small **magnetic field**. *Include how recent studies by the **Galileo probe** indicate that there is probably a **salty ocean** beneath its **crust**.*

_____
_____
_____
_____
_____

5) Hypothesize as to the reasons why the yellowy-orange Galilean moon **Io** is the most **volcanic object** in the solar system.

_____
_____
_____
_____
_____
_____
_____

6) Analyze the reasons why the **Galilean Moons** are responsible for the tremendous donut-shaped **radiation belt** around **Jupiter**.

_____
_____
_____
_____

7) Compare the lines of the **magnetically charged particle flow** dancing around **Ganymede** and **Io** to the **solar wind-driven** neon-colored **aurorae** on **Earth**.

_____
_____
_____
_____

8) Briefly analyze how images of the **surface** of the moon **Europa** show that liquid **water** (or a mixture of slushy ice) is located near its **surface** and under its **solid crust** of **ice**.

_____
_____
_____
_____

9) Clarify how the presence of **silicon rock** on the sea floor of **Europa** could make the **subsurface water** of the moon **salty** and even contain the building blocks for life.

_____
_____
_____
_____
_____
_____

10) Describe the reasons why many **astrobiologists** believe that life outside of **Earth** (if it exists) on other planets and moons would most likely be found in the circumstellar habitable zone of a **star system**. _Connect how the moon **Europa** breaks those rules._

_____
_____
_____
_____
_____
_____

11) Hypothesize as to why many of the distant moons of **Jupiter** are not only **tidally locked** to **Jupiter** but also circle the planet in **retrograde orbits**.

_____
_____
_____
_____
_____

Name_____
Period_____
Date_____

Saturn: Crash Course Astronomy #18

1) Describe the composition of **Saturn** and explain how that affects its **density**.

_____
_____
_____
_____
_____

2) Assess the reasons why the **rapid rotation** and **low density** of **Saturn** is most likely responsible for its **oblate** appearance.

_____
_____
_____
_____

3) Compare the enormous **hexagonal vortex** located on the **north pole** of **Saturn** to the **jet stream** that naturally occurs in the **upper latitudes** on **Earth**.

_____
_____
_____
_____

4) Analyze the composition of the **rings of Saturn**. *Include how the rings are only 10 meters thick.*

_____
_____
_____
_____
_____

5) Hypothesize as to the formation of the **rings of Saturn**.
   A) Hypothesis #1

_____
_____
_____
_____

   B) Hypothesis #2

_____
_____
_____
_____
_____

6) Differentiate between the **three** different **rings** of **Saturn**: A, B, C.
   a) **A Ring**

   _____
   _____
   _____

   b) **B Ring**

   _____
   _____
   _____

   c) **C Ring**

   _____
   _____
   _____

7) Clarify how **resonance** in **celestial mechanics** between the moons **Prometheus** and **Pandora** keep the **F Ring** of **Saturn** intact. _Explain how other **orbiting bodies** inside its rings exert a **regular and periodic gravitational influence** upon each other._

   _____
   _____
   _____
   _____
   _____

8) Analyze the composition of the **atmosphere** on **Titan** to draw conclusions as to why its **atmosphere** is so thick.

   _____
   _____
   _____
   _____

9) Compare **cryovolcanoes** on **Titan** to **volcanos** on **Earth**, and then hypothesize as to whether or not YOU think **Titan** has an **underground** super **salty ocean** of **water**.

   _____
   _____
   _____
   _____
   _____

10) Analyze how frigid temperatures on **Titan** prevent **liquid water** from flowing on its **surface**, yet Titan does have **rivers** and **lakes** of **liquid methane (CH_4)** that flow upon its **surface**.

   _____
   _____
   _____
   _____
   _____
   _____

11) Differentiate how **weather** on the moon **Titan** is driven by **methane (CH₄)** rather than by the **weather** on **Earth** which is driven by **water (H₂O)**.

_____
_____
_____
_____
_____
_____
_____

12) Do YOU think **life** could exist on **Titan**? _Why or why not? Defend your answer._

_____
_____
_____
_____
_____
_____

13) Explain how the **subsurface ocean** on the Saturnian moon **Enceladus** is kept in a **liquid state** even though it is located outside of the **habitable zone**.

_____
_____
_____
_____
_____
_____
_____

14) Do YOU think life could exist on **Enceladus**? _Why or why not? Defend your answer._

_____
_____
_____
_____
_____
_____

15) Hypothesize as to why the moons of **Saturn** orbit the planet in **retrograde**.

_____
_____
_____
_____

16) Evaluate why **Saturn** is considered to be the crown jewel of the solar system.

_____
_____
_____
_____

Uranus & Neptune: Crash Course Astronomy #19

1) Clarify how ice giant **Uranus** got its name after its discovery by musician and astronomer William Hershel (1738-1822) after he initially named it **George's Star** in 1781.

_____

_____

_____

2) Analyze how scientific models of **Uranus** as well as a bit of **physics** have determined how it is made up of three general layers: a) Small **rocky core** b) **mantle - water (H_2O)**, **ammonia (NH_3), methane (CH_4)**, and its c) **atmosphere** – (NH_3), (CH_4), and **hydrogen sulfide (H2S)**.

_____

_____

_____

_____

3) Explain how it might be possible to rain diamonds on **Uranus** due to its intense pressure.

_____

_____

_____

4) Describe the **composition** of the layers of **icy clouds of Uranus** as well as its super cold **atmosphere** of (NH_3), (CH_4), and **(H2S)**.

_____

_____

_____

5) Hypothesize a scenario that explains why the weird **tilt** on **Uranus** has flipped the planet so that it is tilted 98 ° (**degrees**) sideways.

_____

_____

_____

6) Hypothesize as to the reasons why **Uranus** has a set of very skinny **rings.**

_____

_____

_____

_____

_____

7) Compare the similarities between **Uranus** and **Neptune**.

_____

_____

_____

_____

8) Connect the reasons why scientists know that **Neptune** is a lot **denser** than **Uranus**.

_____

_____

_____

_____

9) Assess the reasons why super strong **wind speeds** (some over 2,000 kph) causes **clouds on Neptune** to appear whipped.

_____

_____

_____

_____

10) Compare the vast **internal oceans** and **magnetic fields** of **Uranus** to **Neptune**.
   a) **Uranus** - **magnetic field** crooked, off center, messy

   _____

   _____

   _____

   b) **Neptune** - asymmetric, stretchy arms

   _____

   _____

   _____

11) Compare the rings of **Uranus** to **Neptune**.
   a) **Rings of Uranus**

   _____

   _____

   _____

   b) **Rings of Neptune**

   _____

   _____

   _____

12) Hypothesize as to the reasons why the **orbit** of the moon **Triton** is in **retrograde**.

_____

_____

13) Analyze the surface composition of the Neptunian moon **Triton**, which consists of mostly **frozen nitrogen (N)** along with a mostly **water-ice** crust.

_____

_____

_____

_____

Asteroids: Crash Course Astronomy #20

1) Explain why massive objects such as **Ceres** are considered **asteroids** and not **planets**. *****Ceres** is currently classified as both an **asteroid** and a **dwarf planet**. It is located between the orbits of **Mars** and **Jupiter**.*

_____

_____

_____

_____

2) Hypothesize as to why the composition of **main asteroid belt** are mostly rock and stone with a small number made up of **iron** and **nickel**, and others a mix of the two.

_____

_____

_____

_____

3) Hypothesize as to how the gravitational interactions of **Jupiter** created the desert-like **Kirkwood gaps** in the **main asteroid belt** between **Mars** and **Jupiter**.

_____

_____

_____

_____

4) Speculate as to the role of **salts** in promoting long-term **geological activity** in both the **chemical** and **physical evolution** of ice-rich bodies such as found on the **asteroid Ceres**.

_____

_____

_____

_____

_____

5) Describe the **shape** and analyze the **composition** of the **asteroid Vesta**.

_____

_____

_____

_____

6) Explain why YOU think many **asteroids** are **binary** with the two bodies circling in **orbit** around each other.

_____

_____

_____

_____

7) Analyze the reasons why the **asteroid Itokawa** is a good example of an **asteroid** that is a loosely bound **rubble pile** (i.e. rocks attracted to itself by its own **gravity**).

_____
_____
_____
_____

8) Draw conclusions as to why a **main asteroid belt** exists inside of the solar system.

_____
_____
_____
_____

9) Differentiate between the orbits of **Mars-crossing asteroids**, **Apollo asteroids**, and **Aten asteroids**:
   a) **Mars-crossing asteroids** - cross orbit of **Mars**

   _____
   _____
   _____

   b) **Apollo asteroids** - NEOs (**Near Earth Objects**), cross orbit of **Earth**, orbit more than 1.017 **AU** (**Astronomical Units**)

   _____
   _____
   _____

   c) **Aten asteroids** – **NEOs**, cross orbit of **Earth**, orbit less than 1.017 **AU**

   _____
   _____
   _____

10) Explain how astronomers learn about **asteroids** that can potentially hit **Earth**.

_____
_____
_____

11) Clarify why **asteroids** inhabiting the **Lagrange Points** of **Earth**, and of other planets such as **Jupiter**, have been named **Trojan asteroids.**

_____
_____
_____
_____

12) Explain why **asteroids** that orbit along the same path as **Earth** can stay relatively near the **Earth** in space but don't really orbit **Earth**. _Connect why YOU think that some people call these objects other **moons of Earth**._

_____
_____
_____
_____

Name_____
Period_____
Date_____

Comets: Crash Course Astronomy #21

1) Briefly analyze **Haley's comet** as seen and interpreted on the **Bayeux Tapestry** which commemorated the **Norman Conquest** of **England** at the **Battle of Hastings** in **1066**.

   _____
   _____
   _____
   _____

2) Differentiate between **comets** and **asteroids**.

   _____
   _____
   _____
   _____

3) Analyze how gasses such as **carbon dioxide (CO_2)**, **carbon monoxide (CO)**, **methane (CH_4)**, and **ammonia (NH_3)** along with gravel and dust make up much of the **composition** of a **comet** giving many **comets** the universal name of **dirty snowballs**.

   _____
   _____
   _____
   _____

4) Draw conclusions as to why the **ice** of a **comet** that gets too close to the **Sun** turns directly into a gas by a process called **sublimation**.

   _____
   _____
   _____
   _____

5) Describe the composition of a **comet**, such as **Haley's comet**, being sure to specify the differences between the **nucleus**, the **coma**, and the **tail**.

   _____
   _____
   _____
   _____

6) Clarify how **gas** and **dust** emitted by a **comet** create two different types of **comet tails**.

   _____
   _____
   _____
   _____
   _____
   _____
   _____

7) Cite evidence as to why people on **Earth** were afraid of passing through the **tail of Haley's Comet** in 1910. *Include why the gas **cyanogen (CN)$_2$** played a role in this fear.*

_____

_____

_____

8) Differentiate between **short-period comets** and **long-period comets**.
   a) **Short-period comets** - same **plane** as planets (e.g. **Haley's Comet**)

   _____

   _____

   b) **Long-period comets** - tilted orbits – anywhere in sky (e.g. **Oumuamua**)

   _____

   _____

9) Explain why **comets** do NOT just evaporate when they get close to the **Sun**.

_____

_____

_____

10) Hypothesize about the formation of the **Oort Cloud** where chunks of **dirty ice** are in **orbit** around the **Sun** in an environment where it is perpetually cold far out in **space**.

_____

_____

_____

11) Describe the **surface** of **comet 67P/Churyumov-Gerasimenko** as seen up close by the **Rosetta orbiter** mission and its lander **Philae** in 2014.

_____

_____

_____

12) Defend the hypothesis that **comets** and **asteroids** may have brought a significant amount of **water** to **Earth** billions of years ago not long after the planet had formed.

_____

_____

_____

13) Briefly describe what happened when the **Stardust space probe** passed through the **coma** of **Comet Wild 2** during its attempt to collect **stardust** to return to **Earth**.

_____

_____

_____

The Oort Cloud: Crash Course Astronomy #22

1) Explain the proposal of the **Nice model** where the **migration** of the **giant planets** to their current position was caused by the trillions of **icy objects** that interacted with them in the **solar system** throughout time.

_____
_____
_____
_____

2) Connect the reasons why **Neptune** would have the biggest **gravitational effect** on the **large ice balls** in the **solar system**.

_____
_____
_____
_____

3) Analyze the hypothesis which postulates that during the **Late Heavy Bombardment** (4.1-3.8 billion years ago) **failed planets** and **smaller asteroids** slammed into **larger worlds** after the birth of the **solar system**.

_____
_____
_____
_____

4) Explain why YOU think that **Kuiper Belt objects** (**KBOs**) have **stable orbits**. *Include how they reside in a **circumstellar disk** 4.5-7.5 billion kilometers from the **Sun**.*

_____
_____
_____
_____
_____

5) Evaluate the reasons why it is believed that the **Oort Cloud** is the origin of **long period comets**, and the **scattered disc** is the origin of **short period comets**.

_____
_____
_____
_____
_____
_____
_____

6) Compare the **orbit** of **Neptune** to that of the **trans-Neptunian object** and **dwarf planet Pluto**.

_____
_____
_____
_____
_____
_____
_____

7) Explain how the specific 2:3 **mean-motion orbital resonance** of **Plutinos** interacting with the planet **Neptune** aided in the survival of these **trans-Neptunian objects**.

_____
_____
_____
_____
_____
_____
_____

8) Explain how **Pluto** and its **moons** orbit around a **center of gravity (CG)** rather than **Pluto**.

_____
_____
_____
_____
_____
_____

9) Clarify the reasons why **Pluto** is considered the king of the **Kuiper Belt objects**.

_____
_____
_____
_____
_____

10) Evaluate how scientists have calculated that there should be about **6 billion Oort Cloud objects** that have been left over after the formation of the **solar system**.

_____
_____
_____
_____
_____
_____
_____
_____

Meteors: Crash Course Astronomy #23

1) Clarify what YOU are actually seeing when a **shooting star** zips across the sky.

_____

_____

_____

2) Analyze what happens to a **meteoroid** when it enters the **atmosphere** of **Earth**.

_____

_____

_____

_____

3) Compare the differences between a **meteoroid**, a **meteor**, and a **meteorite**:

   a) **Meteoroid**

   _____

   b) **Meteor**

   _____

   c) **Meteorite**

   _____

4) Explain the formula for the **kinetic energy of an object** $\left(E_k = \frac{1}{2}\, mv^2\right)$ which determines how fast an object is moving.

_____

_____

_____

_____

5) Clarify why most **meteorites**, that are seen streaking though the sky, are only as big as of a grain of sand. _Include the role **kinetic energy** plays in illuminating a **meteor**._

_____

_____

_____

_____

6) Briefly analyze the reasons why a far bigger contributor to the **heat** of a **meteoroid** is **compression** which occurs as a result of its **gas** as it heats up to **hypersonic** speeds.

_____

_____

_____

_____

7) Connect why the streak behind a **meteor** is called a **train**.

_____

_____

_____

8) Explain the reasons why **meteoroids** that travel in packs cause **meteor showers** as they hit the **Earth**. *Include how they initially come from comets not asteroids.*

_____

_____

_____

9) Clarify the **perspective effect** that occurs to viewers on **Earth** during a **meteor shower**.

_____

_____

_____

10) Explain why **meteor showers** that come from the direction of a particular **constellation** are **annual** events. *Include why its radiant determines the name of the meteor shower.*

_____

_____

_____

_____

11) Explain the reasons why a **bolide**, or fireball, is an extremely bright **meteor**.

_____

_____

_____

12) Differentiate between the three different categories of **meteorites**:
   a) **Stony**

   _____

   _____

   b) **Iron**

   _____

   _____

   c) **Stony iron**

   _____

   _____

13) Compare different types of **stony meteorites** called **chondrites** and **achondrites**:
   a) **Chondrites**

   _____

   _____

   _____

   b) **Achondrites**

   _____

   _____

   _____

14) Describe what happened in the **sky** above the Russian city of **Chelyabinsk** in 2013.

_____

_____

_____

_____

Light: Crash Course Astronomy #24

1) Clarify how **light** is a form of **energy** made up of **electromagnetic radiation** with a particular **wavelength** that can be both seen and detected by a human eye.

_____
_____
_____
_____
_____
_____

2) Compare the **wavelength** of **light** to waves in the **ocean**. *Explain how the **energy** of that light is tied to its **wavelength**.*

_____
_____
_____
_____

3) Compare the differences in **wavelength crests** between the color **red** and **violet**. *Include where other colors fit into this **spectrum**.*

_____
_____
_____
_____

4) Differentiate between the parts of the **electromagnetic spectrum**, aka the **visible spectrum** and the **invisible spectrum**. (*e.g. The **invisible spectrum** includes **ultraviolet light, x-ray,** and **gamma rays** on the right side of the **spectrum** and **infrared, microwave,** and **radio** on the left.*)

_____
_____
_____
_____
_____
_____
_____

5) Compare the differences in **light** emitted by **hotter objects** versus **cooler objects** in the **solar system**.

_____
_____
_____

6) Briefly describe the three **subatomic particles** that make up an **atom**:

   a) **Protons**

   _____

   _____

   b) **Electrons**

   _____

   _____

   c) **Neutrons**

   _____

   _____

7) Differentiate between the main components that distinguish the **atoms** of one **element** from another.

   _____

   _____

   _____

8) Analyze the reasons why the **energy** of an **electron** determines where it can occupy specific volumes of **space** around a **nucleus**. _Describe these reasons in concrete terms using the analogy of an **energy staircase**._

   _____

   _____

   _____

   _____

   _____

   _____

   _____

9) Explain the reasons why **electrons** that exist in a **higher energy state** are better equipped to give off **energy** to drop to a **lower energy state**.

   _____

   _____

   _____

   _____

10) Clarify the reasons why a **hydrogen electron** jumps up or down an **energy staircase**. _Include why it absorbs or emits a very specific color of **light** that is different from a **helium electron** or a **calcium electron**._

   _____

   _____

   _____

   _____

11) Draw conclusions as to how **astronomers** can determine what an **object** in the **universe** is made of even if they are not able to physically touch that **object** in **space**.

   _____

   _____

   _____

   _____

Distances: Crash Course Astronomy #25

1) Cite evidence as to how the **Ancient Greeks** knew that the world was round.

_____
_____
_____

2) Assess how Greek philosopher Eratosthenes (276-194 BCE) figured out the size of the **Earth** to be around 40,000 kilometers using his knowledge of the **summer solstice**.

_____
_____
_____
_____

3) Explain how astronomers are able to calculate the **distance** to the **Moon** during a **lunar eclipse**.

_____
_____
_____

4) Explain how the innovative methods used by Eratosthenes to calculate **distances** between objects in the sky were later used by great thinkers such as Hipparchus (190-120 BCE) and Ptolemy (100-170 CE) to get more **accurate sizes** and **distances**.

_____
_____
_____
_____

5) Clarify how Johannes Kepler (1571-1630) and Isaac Newton (1642-1727) laid the mathematical groundwork of **planetary orbits** that in turn made it possible to get the **distances** to all the known **planets** in the **solar system**.

_____
_____
_____

6) Cite evidence as to how the **planetary transits** of **Mercury** and **Venus** crossing the face of the **Sun** could be used to generate numbers to plug into **orbital equations** in order to calculate the length of an **astronomical unit** (AU), or distance of the **Earth** to the **Sun**. *Include why initially there were errors in these timing measurements*.

_____
_____
_____
_____
_____

7) Analyze how knowledge of the **speed of light**, as well as the bouncing of **radio signals** off of the planet **Venus** in the 1960s, gave the correct mathematical number for the **astronomical unit** nailing it down at an **average distance** of 149,597,870.7 kilometers.

_____
_____
_____
_____

8) Explain why knowing the correct number of the **astronomical unit** meant that not only could **astronomers** predict the motions of the **planets, moons, comets,** and **asteroids** but now **probes** could be sent into **space** to explore these strange new worlds up close.

_____
_____
_____
_____

9) Explain how knowledge of the **astronomical unit** unlocked the **distances** to the **stars** due to an understanding of **depth perception** called a **parallax**.

_____
_____
_____
_____

10) Assess how knowledge of the **Earth** orbiting the **Sun** gave **astronomers** the ability to create a **parallax angle** to measure the **distance** from the **Earth** to the **stars** once the size of the **orbit** of the **Earth** had been calculated. _Include the problems with this._ _*Earth's orbit is about 300 million kilometers._

_____
_____
_____
_____

11) Explain how the understanding of the distance **light** travels in a year, aka a **light year** (10 trillion kilometers a year), was needed for a better understanding of the **distance** to **far away stars** such as **61 Cygni**.

_____
_____
_____

12) Explain how **spectroscopy** has allowed **astronomers** to calculate the **distances to nearby stars** using that **spectrographic information** to plug in numbers into the **equation**.

_____
_____
_____

Stars: Crash Course Astronomy #26

1) Analyze the reasons why some **stars** are **brighter** than others in the sky due to both their **distance** as well as their **different colors** of white, red, orange, and blue.

_____

_____

_____

2) Clarify how the science of **spectroscopy** vastly improved interpreting **stellar spectra** by dividing incoming **light** from an **object** into **individual colors** or **wavelengths**.

_____

_____

_____

_____

3) Explain how **spectroscopist** Annie Jump Cannon (1863-1941) created a new system of **star classification** by classifying **stars** dependent on the **absorption lines** in their **stellar spectra** in the late 1800s. *Include the contributions of Max Planck (1858-1947) that the temperature of a star affects its* **observable star light.**

_____

_____

_____

_____

4) Draw conclusions as to how physicist Meghnad Saha (1893-1956) solved how **atoms** give off **light** at different **temperatures** by clarifying which **primary atoms** make up certain **stars**. *Include the role Cecelia Payne-Gaposchkin (1900-1979) played in putting all these pieces together, and by doing so unlocked the* **secrets of the stars.**

_____

_____

_____

5) Explain how the **mnemonic device**, "Oh Be A Fine Guy, Kiss Me," or "Oh Be a Fine Girl Kiss Me," is used to categorize the **spectrum of stars** where each **star** falls in the **heat order classification system**. *CHALLENGE - Design your own **mnemonic device** to remember the even **cooler stars** discovered in the last few decades of **L,T**, and **Y.***

_____

_____

_____

_____

_____

6) Analyze the **classification system** where the yellow-orange **Sun** is a common **G-type main-sequence (G2V)** type star with about a 5500° surface temperature, whereas **G1V** star **47 Ursae Majoris** is slightly hotter, and **G3V** star **16 Cygni B** is slightly cooler.

_____
_____
_____
_____
_____

7) Connect why **Sirius**, the brightest star in the night sky, is much hotter than the **Sun** and classified as an **A0 star**, whereas **Betelgeuse** is red and cool and classified as an **M2 star**.

_____
_____
_____
_____
_____

8) Compare how a lot of **physical characteristics of a star** are related such as how its **luminosity** depends both on its **size** and **temperature**. _Include the reasons why knowing the **temperature and distance to stars** means knowing the **stars** themselves._

_____
_____
_____
_____

9) Analyze the origins of the **HR Diagram** and explain how it functions. _Connect as to why it is sometimes called the **single most important graph** of all of **astronomy**._

_____
_____
_____
_____

10) Differentiate between how various types of **stars**, such as **low mass stars** and **high mass stars**, generate energy when fusing **hydrogen** into **helium** in their **cores**. _Include how this difference explains their position in the **main sequence**._

_____
_____
_____
_____
_____
_____

11) Determine how **stars** can **change position** on the **HR Diagram**. _Include how the **age of a star** plays a significant factor in the **evolution of a star**._

_____
_____
_____
_____

Exoplanets: Crash Course Astronomy #27

1) Explain why finding **exoplanets** is not an easy feat to accomplish in the **night sky** even if there is probably an abundance of them.

_____
_____
_____
_____

2) Create a scenario using **reflexive motion** to explain how one might be able to **indirectly observe** a **planet** around another **star** if it cannot be seen directly with modern technology.

_____
_____
_____
_____

3) Clarify how **radio astronomers** Aleksander Wolszczan (1946-   ) and Dale Frail (1961-   ) changed the **contemporary view** of the universe with their discovery of **two planets** orbiting the **pulsar** PSR 1257+12.

_____
_____
_____
_____

4) Analyze how Michel Mayor (1942-   ) and Didier Queloz (1966-   ) used the **doppler shift** to detect the **transit** of an **exoplanet (51 Pegasi B)** around **Sun-like** star **51 Pegasi** in the **constellation** of **Pegasus** around 50 light years away from **Earth**.

_____
_____
_____
_____

5) Clarify why the amount of **doppler shift** by a **star** is related to the **mass** of its **planet**.

_____
_____
_____

6) Evaluate the problem **astronomers** had rationalizing the existence of the **gas giant exoplanet 51 Pegasi B** (later dubbed a **hot Jupiter**) due to its extremely short **orbital period** of only 4.23 days (orbiting at 8 million kilometers) around its host star **51 Pegasi**.

_____
_____
_____
_____

7) Hypothesize as to why **Jupiter** stayed relatively distant in its **orbital motion** remaining around 473 million miles away from the **Sun**. *Include the role of the planet Saturn.*

_____

_____

_____

_____

8) Clarify how the high **volume** and **mass** of the planet HD 209458 b, aka **Osiris**, just happened to be the first independent observation of an **exoplanet** as well as a marker in understanding the **size** and **density** of an **exoplanet**.

_____

_____

_____

_____

9) Briefly compare the **density** of **Jupiter** to the **density** of **Earth**.

_____

_____

_____

10) Explain how the **Kepler space telescope** found 2,662 **exoplanets** in its decade-long lifetime after being launched into an Earth-trailing **heliocentric orbit** in 2009.

_____

_____

_____

_____

11) Clarify if it has ever been possible to take a **direct photograph** of an **exoplanet**.

_____

_____

_____

_____

12) Clarify if any rocky **Earth-like planets** have been found in the **habitable zone of a star**, aka its **Goldilocks zone**. *Analyze how the planets around TRAPPIST-1 have been confirmed to be rocky planets with possible liquid water on their surface.*

_____

_____

_____

_____

_____

_____

13) Hypothesize the problems a planet has when trying to retain an **Earth-like atmosphere**.

_____

_____

_____

_____

_____

Name_____
Period_____
Date_____

Brown Dwarfs: Crash Course Astronomy #28

1) Explain why an object needs to have at least **75 times** the **mass of Jupiter** in order to become a **star**.

_____

_____

_____

_____

2) Analyze what would happen to an **object** that forms like a **star** but does NOT have at least **75 times** the **mass of Jupiter**.

_____

_____

_____

_____

3) Clarify the roll that Jill Tarter (1944-   ) had in giving **brown dwarfs** their name.

_____

_____

_____

_____

4) Clarify the reasons why **L-stars** are NOT **brown dwarfs**.

_____

_____

_____

5) Explain why **brown dwarfs** lighter than about 65 times the **mass of Jupiter** would NOT fuse **lithium** in their **cores**. *Include how this provided **Brown dwarf hunters** with a test to distinguish **brown dwarfs** from regular **stars**.*

_____

_____

_____

_____

6) Differentiate between the **atmospheres** of **brown dwarfs** Teide 1, Gliese 229b, and Wise 1828+2650:
a) Teide 1 – hot – **L-dwarfs**

_____

_____

_____

b) Gliese 229b – cool - **T-dwarfs**

_____

_____

_____

c) Wise 1828+2650 – very cool – **Y dwarfs**

_____

_____

_____

7) Describe the mission of NASA's **Wide-Infrared Survey Explorer (WISE)** that was designed to scan the entire sky with **infrared light** and launched in 2009. *Include the types of objects it found.*

_____

_____

_____

_____

8) *Challenge - Construct your OWN mnemonic sentence to remember all the classes of stars that include brown dwarfs: **O,B,A,F,G,K,M,L,T,Y**. *Here is a mnemonic provided for you: Oh Boy About Four Green Kangaroos Might Locomotion-hop Toward You.*

_____

_____

_____

_____

9) Explain why the coolest of the **brown dwarfs**, that aren't completely black, would emit the color **magenta**. *Include why WISE would see these objects as the color green.*

_____

_____

_____

_____

10) Explain why **brown dwarf** stars would only get denser and not any bigger if more mass was added to them. *Include the differences between a small brown dwarf and a really big planet.*

_____

_____

_____

_____

11) Explain the reasons why **brown dwarfs** that are more massive than about 13 times the mass of **Jupiter** can fuse **deuterium** (^{2}H) in their **cores** even though they are NOT considered true stars.

_____

_____

_____

_____

12) How would YOU describe a **brown dwarf** to a person who did not know what one was?

_____

_____

_____

_____

Low Mass Stars: Crash Course Astronomy #29

1) Briefly analyze how a **star** makes **energy** in its **core** by fusing **hydrogen** into **helium**. *Include how its **power** comes from that **released energy** dependent upon its **pressure**.*

_____

_____

_____

_____

2) Differentiate between **low mass stars** and **high mass stars**. Include how the **energy released** is dependent upon the **pressure** in its **core**. *$^4H => He + E$*
   * *$E=mc^2$ (**Energy** = **mass** x **speed of light** squared)*
   A) **Low mass stars**

   _____

   _____

   _____

   _____

   B) **High mass stars = 8X mass of the Sun**

   _____

   _____

   _____

   _____

3) Explain the reasons why the **lowest mass stars**, known as **cool red dwarfs**, have very long lifespans. *Include how **convection** works to power a cool **red dwarf star**.*

   _____

   _____

   _____

   _____

4) Analyze how stars like the **Sun** which have bigger, hotter, and denser **cores** than **red dwarf stars**, are NOT **conducive** to **convection** in their **cores** but are **conducive** to **convection** inside their **convective zones**.

   _____

   _____

   _____

5) Clarify how the **core** of the **Sun** gets hotter each day. *Include why the **luminosity** of the **Sun** is 40% brighter today than when it was born.*

   _____

   _____

   _____

   _____

6) Explain why the **Sun** will eventually run out of **hydrogen** to fuse in its **core**. *Clarify why it is more difficult to fuse the atoms of **helium (He)**, **oxygen (O)**, and **neon (N)** in its **core**.*

_____
_____
_____
_____

7) Draw conclusions as to why the outer layers of the **Sun** will continue to expand and cause the **Sun** to grow bigger over its lifetime as well as causing its **color** to change. *Include how the **Sun** will eventually become a **red giant star** and increase its **luminosity** 200 times.*

_____
_____
_____
_____
_____
_____

8) Describe the reasons why **gravity** on the surface of the **Sun** as a **red giant star** will diminish. *Explain how the **Sun** will shed more **mass** from its **surface** than it does today.*

_____
_____
_____
_____
_____
_____

9) Actualize a logical argument that explains why the **Sun** will turn from red to orange and back to red once **helium fusion** occurs in the **core** of the **Sun**.

_____
_____
_____
_____

10) Determine why the **Sun** will become a **white dwarf** at the end of its life after it has converted all of its **helium (He)** into **carbon (C)** in its **core**.

_____
_____
_____
_____

11) ***Bonus Question***
Develop a **thesis** as to how a **Dyson Sphere** built around the **Sun** to collect **energy** would solve all the **energy problems** currently faced on **Earth** now and in the far **future**.

_____
_____
_____
_____

White Dwarfs & Planetary Nebulae: Crash Course Astronomy #30

1) Explain what will happen to the **Sun** when it runs out of **hydrogen (H)** and **helium (He)** to fuse in its **core** since it will NOT have enough **pressure** to squeeze **carbon nuclei** together in around 8 billion years from now.

_____
_____
_____
_____
_____

2) Clarify how **electron degeneracy pressure**, a result of **quantum mechanics** surrounding **The Pauli Exclusion Principle**, describes how **sub-atomic particles** behave on teeny tiny scales.

_____
_____
_____
_____
_____
_____

3) Briefly analyze the **characteristics** of a **white dwarf star**.

_____
_____
_____
_____
_____
_____

4) Analyze both the **temperature** and the **glow** on the **surface** of a **white dwarf star**.

_____
_____
_____
_____
_____

5) Describe how **planetary nebulae** are related to **white dwarf stars**.

_____
_____
_____
_____
_____
_____
_____
_____

6) Clarify as to how **planetary nebulae** come in all sorts of fantastic shapes and are not just circular **planetary nebulae**. *Include how the **centrifugal force** of a **binary star system** plays a role in creating these fantastic **planetary nebulae shapes**.*

_____
_____
_____
_____
_____
_____

7) Hypothesize as to how **planets** swallowed up by their **stars** during their **red giant star** phase could be responsible for the **shapes of planetaries**.

_____
_____
_____
_____
_____

8) Analyze how the central part of the **white dwarf star** exciting the **gas** surrounding it plays a key role in causing the **glow** in a **planetary nebula**.

_____
_____
_____
_____
_____
_____

9) Explain how it is possible to understand about **stellar evolution** by studying how **stars** meet their **death**.

_____
_____
_____
_____
_____

10) Connect the reasons why the **planetary nebulae phase** is relatively short-lived.

_____
_____
_____
_____

11) Hypothesize as to what will happen to the **Sun** in the **distant future**. *Include the reasons why YOU think it will or will NOT have a **planetary nebula** that is visible in the **sky**.*

_____
_____
_____
_____

High Mass Stars: Crash Course Astronomy #31

1) Briefly analyze the reasons why **stars** are in a constant struggle between the **collapsing power of gravity** and their **internal heat** trying to **inflate** them. *Include a brief analysis of what will happen to our Sun in its twilight years.*

_____
_____
_____
_____
_____

2) Give an overview as to what is going on in the **core** of a **star** in order for **energy** to be released and **new elements** to be created.

_____
_____
_____

3) Explain what will happen to a **star** if it has more than **8 times** the **mass** of the **Sun**. *Include how hot it needs to be in order for a star to fuse carbon.*

_____
_____
_____
_____

4) Illustrate the **characteristics** of **red hypergiant stars** such as **VY Canis Majoris** and **UY Scuti** in comparison to a **yellow star** like the **Sun**.

_____
_____
_____
_____
_____
_____

5) Explain the reasons why **red supergiant stars** like **Betelgeuse** in the **constellation Orion** do NOT last longer than yellow stars like the **Sun**. *Research and report on the reasons why many people thought Betelgeuse was going to explode in a supernova in 2019.*

_____
_____
_____
_____
_____
_____
_____

6) Analyze the reasons why a **massive star** fusing **silicon** into **iron** spells trouble for that **star** and makes it a ticking time bomb due to what is happening in its **collapsing core**.

_____
_____
_____
_____
_____

7) Assess how the **size** of a **star** determines whether or not a **neutron star** forms in **space** rather than a **supernova** or **black hole** when a **large star collapses** inward upon itself.

_____
_____
_____
_____
_____

8) Deconstruct how **sub-atomic energy particles** called **neutrinos** are created in a **supernova**.

_____
_____
_____
_____

9) Describe some of the fantastic **shapes** formed by a **supernova remnant**. _Include why the_ **_Crab Nebula_**_, which formed from the_ **_supernova_** _of a_ **_star_** _in 1054 (_**_SN 1054_**_), is so famous. *Hint-_ **_East-West Schism_** _(1054); visible in daytime for a couple of weeks_

_____
_____
_____
_____
_____

10) Draw conclusions as to why **Earth** probably will NOT be hit by a **supernova**.

_____
_____
_____
_____

11) Evaluate how **supernovae** are capable both of **great destruction** as well as being a critical factor for the existence of **life** as we know it on **Earth**.

_____
_____
_____
_____

Neutron Stars: Crash Course Astronomy #32

1) Clarify how **electrons** in a **1.4 solar mass star** react to being squeezed together when its **core** collapses. *Explain why it becomes a **neutron star**.*

_____
_____
_____
_____

2) Explain what happens to **neutrons** in a **star** that has less than **2.8 solar masses**. *Clarify why this **collapse** creates **neutrinos**.*

_____
_____
_____
_____

3) Analyze the reasons why a **neutron star** that is only about 20 kilometers across has such mind-crushing properties. *Clarify how heavy a spoonful of **neutronium** would **weigh**.*

_____
_____
_____
_____

4) Calculate how much YOU would **weigh** if you were able to stand on the **surface** of a **neutron star**, and then explain why you would **weigh** so much. *(*Hint – multiply YOUR weight by 100 billion pounds)*

_____
_____
_____
_____

5) Clarify what would happen to the **spin of a star** which had shrunk to just 20 km across to become a **neutron star**. *Include its new **magnetic properties**.*

_____
_____
_____
_____

6) Explain how Jocelyn Bell (1943-   ) figured out that she had discovered the first known **pulsar** when she observed its **beams** sweep across the **star** rotating like the beams emitted from a lighthouse.

_____
_____
_____
_____

7) Explain how it is now known that there is a **pulsar** at the center of the **Crab Nebula**.

_____
_____
_____
_____
_____

8) Analyze the reasons why über-powerful **magnetars** have a **magnetic field** a quadrillion-times stronger than the **magnetic field** of the **Sun**, and are even 1000 times more powerful than the most super magnetic **neutron stars**.

_____
_____
_____
_____
_____

9) Differentiate between what happens during an **earthquake** and what happens during a **starquake**.
   a) **Earthquake**

_____
_____
_____
_____

   b) **Starquake**

_____
_____
_____
_____

10) Differentiate between a **solar flare** and a **magnetar flare**.
   a) **Solar flare**

_____
_____
_____

   b) **Magnetar flare**

_____
_____
_____

11) Connect how powerful **magnetars** are by analyzing what happened after a huge blast of **X-rays** slammed into some **satellites** orbiting the **Earth** from **magnetar SGR-1806-20** located behind the center of the **Milky Way** some 50,000 light years away in 2004.

_____
_____
_____
_____

Black Holes: Crash Course Astronomy #33

1) Briefly explain what happens when a **star dies** and collapses inward upon itself.
   *Depending on its **mass**, determine whether it becomes a **white dwarf**, a **neutron star**, or
   a **black hole**.*
   A) **white dwarf** < 1.4 **solar mass**

   _____

   _____

   _____

   B) **neutron star** =1.4 >2.8 **solar mass**

   _____

   _____

   _____

   _____

   C) **black hole** > 2.8 **solar mass**

   _____

   _____

   _____

   _____

2) Describe how **escape velocity** plays a role after the **core** of a **high mass star** collapses.

   _____

   _____

   _____

   _____

   _____

   _____

   _____

3) Briefly explain why a **rocket** would need to have an **escape velocity** of **11 km/sec** in
   order to leave the **gravitational pull** of the **Earth**.

   _____

   _____

   _____

   _____

4) Briefly explain why a **rocket** would need to have an **escape velocity** of **600 km/sec** in
   order to leave the **gravitational pull** of the **Sun**.

   _____

   _____

   _____

   _____

   _____

5) Briefly explain why a **rocket** would need to have an **escape velocity** of **150,000 km/sec**, or **half the speed of light**, in order to leave a **neutron star**.

_____
_____
_____
_____
_____

6) Briefly explain why **light** cannot escape from a **black hole**.

_____
_____
_____
_____
_____
_____

7) Analyze what happens to **time** and **space** along the **event horizon** of a **black hole** where the **escape velocity** is at the **speed of light (300,000 km/sec)**.

_____
_____
_____
_____
_____
_____
_____

8) Clarify the reasons why the **Sun** cannot become a **black hole** due to its insufficient **mass**.

_____
_____
_____
_____
_____
_____

9) Assess the reasons why it is possible to **orbit** a **black hole** without falling inside of it.

_____
_____
_____
_____
_____
_____
_____

10) Analyze the reasons why **spaghettification** would occur if a person jumped **feet first** into a **stellar mass black hole**.

_____

_____

_____

_____

_____

11) Differentiate between **stellar mass black holes** and **supermassive black holes**:
   a) **Stellar mass black holes**

   _____

   _____

   _____

   _____

   b) **Supermassive black holes**

   _____

   _____

   _____

   _____

12) Analyze how the **fabric of space** around a **black hole** is **warped by gravity**.

_____

_____

_____

_____

_____

_____

13) Explain how **time** and **space** are two parts of the same thing called **space time**, and that one cannot be affected without affecting the other.

_____

_____

_____

_____

_____

14) Evaluate the reasons why **two people** viewing each other at **different positions** around a **black hole** would perceive an event differently; While one person would see a **gravitational redshift** if that person were viewing a **black hole** right outside its **event horizon, the other** person inside the **inner horizon** and falling into the **black hole** would see a **gravitational blueshift**.

_____

_____

_____

_____

_____

_____

Name_____
Period_____
Date_____

Binary and Multiple Stars: Crash Course Astronomy #34

1) Clarify the reasons why some **stars** that appear to be a **single star** in the sky are actually **multiple stars** very far apart and aptly named **optical double stars**.

_____
_____
_____

2) Analyze the reasons why astronomers call **stars** that physically **orbit** each other **binary stars** rather than **double stars**.

_____
_____
_____

3) Connect the relationship between the stars **Mizar** and **Alcor** which are located around the kink in the **handle of the Big Dipper** inside the **constellation** of **Ursa Major**.

_____
_____
_____

4) Explain how **binary stars** form inside of their original **stellar nursery**.

_____
_____
_____

5) Clarify how the luminous blue star **Sirius** is a **visual binary** as well as the **brightest star** in the night sky with a **visual magnitude** of -1.44. *Include how under an X-Ray telescope, Sirius B (a white dwarf) can be seen.*

_____
_____
_____
_____

6) Analyze how the **size** and **shape** of the **orbit** around a **binary star** can be determined, once the **distance** to that **binary star** has been calculated. *Include how the masses of these binary stars can then be found by using the math and physics of gravity.*

_____
_____
_____
_____

7) Clarify how it is possible to learn everything else about a **star** once the **mass of the star** is known. *Include how this relates to the field of astrophysics.*

_____
_____
_____

8) Connect how **spectroscopy** allows astronomers to understand that a **star** is **binary** (aka **spectroscopic binaries**) by analyzing the **doppler shift** in their **spectrum** even when these **stars** are unable to be split even with the biggest **telescopes** on **Earth**.

_____
_____
_____
_____

9) Analyze the reasons for the **X-ray burst** source **4U 1820-30** which hosts a **neutron star** and a **white dwarf** inside the **core** of **globular cluster NGC 6624**.

_____
_____
_____
_____

10) Explain why the most famous **eclipsing binary** star **Algol** (aka beta Persei) dims for several hours once every three days. _Include the reasons why astronomers know so much about this **binary star system**._

_____
_____
_____
_____

11) Describe the reasons why W Ursae Majoris (**W Uma**) is classified as a **contact binary** in the **northern constellation** of **Ursa Major**. _Include the role **mutual tidal effects** play in fashioning these **stars** into the shape of a double-lobed stellar peanut._

_____
_____
_____
_____

12) Describe how the **Algol paradox** explains why the **eclipsing binary** star **Algol** has evolved into somewhat of a paradoxical situation. *_Include how elements of this **binary star system** seem to have evolved in discord with established theories of **stellar evolution**._

_____
_____
_____
_____
_____
_____

13) Differentiate between a **stellar nova** and a **recurrent nova**. _Include how a **supernova** can happen to a **white dwarf star** even though initially it does NOT have enough **mass**._

_____
_____
_____
_____

Star Clusters: Crash Course Astronomy #35

1) Differentiate between **galactic clusters** and **globular clusters**:
   a) **Galactic clusters (open clusters)** – young

   _____
   _____
   _____
   _____
   _____

   b) **Globular clusters** – spherical shape, well defined **core**, dense, lots of **stars**, old

   _____
   _____
   _____
   _____
   _____

2) Explain how **stars** orbit each other in **galactic (open) clusters**, and draw conclusions as to why the **stars** do NOT all orbit a **center of mass** like **planets** that **loop around a star** on a **flat plane.**

   _____
   _____
   _____
   _____

3) Analyze how astronomers can calculate the **age of star clusters** by determining the **types of stars** inside of an **open cluster.** (e.g. **blue stars, red dwarf stars**)

   _____
   _____
   _____
   _____

4) Connect how **gravity** causes there to be very few **star clusters** older than roughly 50 million years, as a large number of **stars** have been lost due to time.

   _____
   _____
   _____

5) Evaluate what is going on inside of the **open Pleiades star cluster** (the **Seven Sisters**) which is located about 500 light years from **Earth** within the **constellation** of **Taurus.**

   _____
   _____
   _____
   _____

6) Evaluate what is going on inside of the **open Hyades star cluster** which comprises the horns of the **Taurus constellation** about 153 light years from **Earth**.

_____
_____
_____
_____

7) Analyze the reasons why it took **billions of years** after the **birth of the universe** for **stars** like the **Sun** to form.

_____
_____
_____
_____
_____
_____

8) Clarify why all the **stars** in a **globular cluster** are not only the same age but roughly the same distance from **Earth**.

_____
_____
_____
_____

9) Connect the reasons why **globular clusters** tend to look **red**.

_____
_____
_____
_____

10) Hypothesize as to how there can be **blue stars** inside of **globular clusters** if all the stars are the same age.

_____
_____
_____
_____

11) Compare **globular clusters** to the illusion of bees swarming around a hive.

_____
_____
_____
_____

12) Connect how it could be possible to **read by starlight** if the planet you were on were located inside of a **globular cluster**. _Include why this scenario can most likely be found only inside of the realm of **science fiction** and the **imagination** and NOT in reality._

_____
_____
_____
_____

Nebulae: Crash Course Astronomy #36

1) Briefly describe how **nebulae** are really just clouds of **gas and dust** in **space**.

_____

_____

2) Explain how the **Sun** formed from a **nebula** after a **high mass star** died in a **supernova** around 4.6 billion years ago.

_____

_____

_____

_____

3) Analyze how a **cloud of gas** lights up in an **emission nebula** after being blasted by **light** from a nearby **massive star**.

_____

_____

_____

_____

4) Describe the **density and pressure** of an **average nebula** that is **light years** in **diameter** by comparing its **atoms** to the **air density and pressure** found on the planet **Earth**.

_____

_____

_____

_____

5) Differentiate between **emission nebulae** and **reflection nebulae**:
   A) **Emission nebulae** – own light glow

   _____

   _____

   _____

   _____

   B) **Reflection nebulae – nearby stars** cause glow, contain tiny grains of dust (**silicates, aluminum oxide, calcium**) and soot (**polycyclic aromatic hydrocarbons**).

   _____

   _____

   _____

   _____

6) Briefly clarify how **stardust**, aka **cosmic dust**, can affect the **visible light** from **stars** even if it does NOT emit **visible light**.

_____

_____

_____

7) Analyze why **red light** can more easily pass-through **dust clouds** than **blue light**. *Defend your answer with an example of the **Maia Nebula dust cloud** surrounding the **Pleiades open star cluster**.*

_____

_____

_____

_____

8) Clarify how **thick dust** is not only very good at absorbing **visible light** but can also **dim a star** considerably if the **dust cloud** is big enough or dense enough.

_____

_____

_____

9) Analyze the effect that the **dark absorption nebula**, inside the **dark interstellar cloud** known as **Barnard 68**, has to observers in the **Solar System** due to its scattering of **blue light** but not **red light**. *Hypothesize as to what might happen to **Barnard 68** in two hundred thousand years.*

_____

_____

_____

_____

10) Differentiate between an **emission nebula** such as the **giant molecular cloud** of the **Orion Nebula** and an **absorption nebula** such as **Coalsack Nebula (Southern Coalsack)**.

_____

_____

_____

_____

11) Describe how **giant molecular clouds** such as those in the **Orion Nebula** can stretch for hundreds of light years to form a cloud of ionized gas called an **emission nebula**.

_____

_____

_____

_____

12) Analyze how the **Eagle Nebula,** aka the **Pillars of Creation**, is actually a **star factory** that has had its shape form because of the **photoevaporation** of **ionized gas** slamming into the much thinner gas of the **interstellar medium** after it was created by a **supernova**.

_____

_____

_____

_____

13) Clarify how **stellar wind** compresses **gas** as **stars** explode and when they are born.

_____

_____

_____

The Milky Way: Crash Course Astronomy #37

1) Connect how Galileo (1564-1642) came to understand that he was looking not at **dust clouds** in the **night sky**, but at a massive grouping of **individual stars** at the **center of the Milky Way Galaxy** when he looked up at the **night sky** through a **telescope** in 1610.

_____
_____
_____
_____

2) Explain how **globular clusters** provided a good clue that there was more going on than just the **stars** that could be seen in the **Milky Way Galaxy**.

_____
_____
_____
_____

3) Describe the structure of the **Milky Way**. *Include how using **light** other than **visible light**, such as found in **radio astronomy**, aids in our understanding of the size and shape of the **Milky Way** as **dust** severely restricts our view from inside the **solar system**.*

_____
_____
_____
_____

4) Analyze how it is known that **massive stars** form inside of **giant nebulae** located in the **spiral arms** of the **Milky Way Galaxy**.

_____
_____
_____
_____

5) Evaluate the reasons for the current placement of **stars** in the **spiral arms** of the **Milky Way Galaxy** by likening them to a traffic jam of cars. *Include why **stars** and **nebulae** closer to the **galactic center** orbit faster.*

_____
_____
_____
_____
_____
_____
_____
_____

6) Explain the reasons why it takes the **Sun** 250 million years to complete one **orbit** around the **Milky Way Galaxy**. *Clarify how many times it has circled it since its birth.*

_____

_____

_____

_____

7) Draw conclusions as to why the **spiral arms** of the **Milky Way Galaxy** appear to be **blue**.

_____

_____

_____

_____

8) Connect how astronomers who study **stellar kinematics** know that the **Sun** is located just off of one of the smaller **Milky Way arms** called the **Orion Arm**.

_____

_____

_____

_____

9) Clarify the reasons for the reddish colored **bar-like shape** at the **center** of the **Milky Way Galaxy** which is estimated to be about 20,000 light years wide.

_____

_____

_____

_____

10) Hypothesize as to the reasons why the **bar** at the **center** of the **Milky Way Galaxy** has a **stellar orbit** which rotates as a **single unit**.

_____

_____

_____

_____

11) Connect how a very massive **black hole (Sagittarius A*)** at the center of the **Milky Way Galaxy** may be responsible for the **formation** and **evolution** of the **Milky Way**.

_____

_____

_____

_____

_____

12) Clarify how astronomers know that the **stars** in the **spherical halo** of the **Milky Way Galaxy** are very old.

_____

_____

_____

_____

_____

Name_____
Period_____
Date_____

Galaxies, part 1: Crash Course Astronomy #38

1) Evaluate how the discovery of **cepheid variables** inside of the **Andromeda Galaxy (M31)** by Edwin Hubble (1889-1953) unlocked the mystery as to whether or not the universe consisted of just the **Milky Way** or if it was just one of many **galaxies** in the universe.

_____

_____

_____

_____

_____

2) Briefly explain how big YOU think the universe must by **calculating** how many **stars** are inside an **average galaxy**.

_____

_____

_____

3) Differentiate between the size and shapes of the **four major types of galaxies: spiral, elliptical, peculiar, and irregular.**

   a) **Spiral – Milky Way –** broad, flat, rotating disks of stars and dust; **Grand Design spirals, flocculent spirals**

   _____

   _____

   _____

   _____

   i)   **Grand Design** -well-organized

   _____

   ii)   **Flocculent spirals** – cotton-like, patchy arms

   _____

   b) Differentiate between the **spiral galaxies** which we see **straight-on** from **Earth** versus **galaxies** that we see **tilted** and at an **angle**.

   i) **Straight-on galaxies**

   _____

   _____

   _____

   ii) **Tilted galaxies**

   _____

   _____

   _____

4) Clarify why the **solar system** is flat.

_____

_____

_____

5) Differentiate between the size and shapes of the **four major types of galaxies**: spiral, **elliptical**, peculiar, and irregular.
    a) **Elliptical** – puffy, **M87,** little gas and dust, little star formation, older stars

    _____

    _____

    _____

        i)   **Dwarf ellipticals**

        _____

        _____

        _____

        ii)   **Large elliptical**

        _____

        _____

        _____

6) Describe what happens when there is a **collision** between **two galaxies**.

_____

_____

_____

_____

_____

7) Analyze what happens to **gas clouds** inside of **galaxies** that collide.

_____

_____

_____

_____

8) Explain what happens when a **big spiral galaxy** collides with a much **smaller galaxy**.

_____

_____

_____

9) Prove how the **Milky Way Galaxy** is currently cannibalizing several smaller **galaxies** such as the **Sagittarius Dwarf Elliptical Galaxy** and **Canis Major Dwarf Galaxy**.

_____

_____

_____

_____

_____

_____

10) Differentiate between the size and shapes of the **four major types of galaxies**: spiral, elliptical, **peculiar**, and irregular.

    a) **Peculiar**

_____

_____

_____

_____

        i)   **Ring Galaxy – Cartwheel galaxy, Hoag's object**

                 Ψ) **Cartwheel galaxy**

_____

_____

                 Z) **Hoag's object**

_____

_____

11) Differentiate between the size and shapes of the **four major types of galaxies**: spiral, elliptical, peculiar, and **irregular**.

    a) **Irregular**

_____

_____

        i)   **Large Magellanic Clouds (LMC) -Southern hemisphere**, 200K light years from **Earth**, high mass star formation

_____

_____

_____

                 Ψ) **30 Doradus (Tarantula Nebula)**

_____

_____

_____

        ii)   **Small Magellanic Clouds (SMC) - Southern hemisphere**

_____

_____

_____

12) Clarify the difference between **active galaxies** and **normal galaxies**.

_____

_____

_____

13) Why do YOU think **galaxies** tend to cluster together in the known universe, with some clusters containing **thousands of galaxies**?

_____

_____

_____

_____

Galaxies, part 2: Crash Course Astronomy #39

1) Explain how the advent of **X-Ray observatories** launched into **space** allow astronomers to identify and classify objects such as **quasars** inside of **active galaxies**.

_____
_____
_____
_____
_____

2) Analyze the reasons behind how astronomers came to the conclusion that all immensely energetic and **active galaxies** have very **massive black holes** at their center.

_____
_____
_____
_____
_____

3) Clarify the theory which explains that all **galaxies** have a **black hole** at their center. *Include how stars can be shredded by the fierce **gravity** of a **black hole** which in turn creates an illuminated **accretion disk**.*

_____
_____
_____
_____
_____
_____
_____

4) Briefly explain how **friction** and **other forces** power **active galaxies** to blast light across the **electromagnetic spectrum**.

_____
_____
_____
_____

5) Explain how **active galaxies** can have **jets** due to their intense **magnetic fields** as well as the **super-fast rotation** of their **accretion disks**.

_____
_____
_____
_____
_____
_____

6) Analyze the reasons why the **accretion disks** of **active galaxies** can look pretty different from each other as seen from the **viewing angle** of **Earth**.

_____
_____
_____
_____
_____

7) Hypothesize as to how a **galactic collision** between the **two galaxies** such as the **Andromeda galaxy** and the **Milky Way galaxy** could turn-on the **black hole** at the center of the **Milky Way galaxy** and change it from a **normal galaxy** into an **active galaxy**.

_____
_____
_____
_____
_____
_____
_____

8) Connect how the **Milky Way** is part of a local group of **galaxies** clumping together.

_____
_____
_____
_____
_____
_____

9) Defend how a **blue shift** in the **light spectrum** of the **Andromeda galaxy** has convinced astronomers that the **Andromeda galaxy** is headed towards the **Milky Way** and will be here in our backyards in approximately 4 billion years.

_____
_____
_____
_____
_____
_____

10) Describe how a new **super galaxy** nicknamed **Milkomeda** will form as a result of the collision between the **Andromeda galaxy** and the **Milky Way galaxy**.

_____
_____
_____
_____
_____
_____

11) Calculate the size of a typical **galaxy cluster** such as the **Virgo cluster** which is located about 50 million light years away in the direction of the **constellation** of **Virgo**. *Defend your answer.*

_____

_____

_____

_____

_____

12) Calculate the size of the **Virgo Supercluster** and the even larger **Laniakea Supercluster** of which the **Milky Way** and the **Andromeda Galaxy** call home. *Defend your answer.*

_____

_____

_____

_____

_____

13) Explain how the **universe** appears almost foamy on the biggest scales and looks kind of like a sponge interspersed with **filaments**.

_____

_____

_____

_____

_____

14) Describe what the **Deep Field** lens of the **Hubble Space Telescope** found when it pointed its **optics** towards a tiny section of the sky that seemed completely empty.

_____

_____

_____

_____

_____

15) Evaluate how it is possible for **astronomers** to calculate the approximate number of **galaxies** in the universe by **extrapolating** the entire sky in order to calculate the **net total** number at **100 billion galaxies**.

_____

_____

_____

_____

_____

_____

Gamma-Ray Bursts: Crash Course Astronomy #40

1) Describe how a flash of **gamma rays** picked up by the **Vela satellites** after it was launched by the US to monitor the **nuclear activity** of the Soviet Union in **space** led to the discovery of a **deep space** event on July 2, 1967.

_____
_____
_____
_____
_____
_____

2) Explain the reasons why it is difficult to generate a **gamma ray**. *Defend your answer with examples.*

_____
_____
_____
_____

3) Clarify why scientists have concluded that **gamma ray bursts** (**GRBs**) are actually coming from **distant galaxies** far away from the **Milky Way**.

_____
_____
_____
_____

4) Evaluate how solving where **GRBs** had originated from was due to a faster response time in capturing the afterglow of a **GRB** before it became invisible.

_____
_____
_____
_____

5) Hypothesize as to what type of object would have to be responsible to create the intensely powerful **GRBs**. *Defend your answer.*

_____
_____
_____

6) Differentiate between a **supernova** blast from a **star** and a **beam of light** from a **GRB**.

_____
_____
_____
_____

7) Describe the type of **GRB** that is produced when the **core** of a **very massive star** collapses to form a **black hole** releasing a pair of columnated **gamma-ray** jets.

_____

_____

_____

_____

8) Clarify the differences between **supernova** and a **hypernova**:
   a) **Supernovae**

_____

_____

_____

   b) **Hypernova**

_____

_____

_____

9) Compare the two kinds of **GRBs**.
   a) **Hypernova** - last longer than 2 seconds

_____

_____

   b) **Neutron Stars** – collision (less than 2 seconds)

_____

_____

10) Analyze the bizarre outcome between the collision of two **neutron stars** that combine together to create a mass **2.8 times** that of the **Sun**.

_____

_____

_____

_____

11) Hypothesize as to what might happen if a **GRB** exploded within **7,000 light years** of **Earth**.

_____

_____

_____

_____

12) Analyze the chances of a **GRB** hitting **Earth** any time in the near future.

_____

_____

_____

13) Describe the success of **NASA's Swift spacecraft** in observing **GRBs** and their afterglow.

_____

_____

_____

Dark Matter: Crash Course Astronomy #41

1) Explain how it is possible for astronomers to not only calculate the **mass** of the **Earth** and the **Sun** but to also to calculate the **mass** of the entire **Milky Way Galaxy** using the **Doppler shift** to calculate its **velocity**.

_____
_____
_____
_____
_____
_____

2) Clarify how the existence of some type of **dark material**, such as **dark matter**, would explain why **stars** farther out in arms of the **Milky Way Galaxy** actually move faster than those closer to its **gravitational center**.

_____
_____
_____
_____
_____
_____
_____

3) Differentiate between the **virial theorem of mechanics** concerning **dark matter** by astronomer Fritz Zwicky (1898-1974), and **galaxy rotation rate** pioneer and astronomer Vera Rubin (1928-2016). *Explain which observation turned out to be more accurate.*
   A) Fritz Zwicky – **viral theorem of mechanics**, coined **dark matter** term

_____
_____
_____
_____
_____
_____

   B) Vera Rubin – **galaxy rotation rate**

_____
_____
_____
_____
_____
_____
_____

4) Evaluate the supposition of **axions** as being a prime contender for cold **dark matter** (which makes up around 85 percent of all **matter** in the universe) by analyzing how **indirect methods** of observation such as **gravitational lensing** can prove its existence.

_____
_____
_____
_____
_____
_____

5) Analyze how scientists have used **gravitational lensing** not only to map out where the **hot gas** was inside of the **Bullet Cluster** but also to measure its **mass** (including the **mass** of **dark matter**) from the even further **galaxies** distorted in the background images.

_____
_____
_____
_____
_____
_____

6) Defend the hypothesis that **dark matter** is made out of **axions** by confirming why **axions** should be surrounding the **galaxy clusters** themselves. _Defend YOUR answer with observations as seen by the **Chandra X-Ray Observatory**._

_____
_____
_____
_____
_____
_____

7) Consider other candidates besides **axions** which might account for **dark matter** by hypothesizing and analyzing other ideas about **composition** of **dark matter**.

_____
_____
_____
_____
_____
_____

8) Evaluate how **dark matter** has had a profound effect on the **universe**.

_____
_____
_____
_____

Name_____
Period_____
Date_____

The Big Bang, Cosmology part 1: Crash Course Astronomy #42

1) Evaluate whether or not YOU agree with the analysis that our universe is basically just a **static universe**, and is where it is and always has been as we see now see it, generally unchanging since the beginning of time. *Defend your answer with examples.*

_____
_____
_____
_____
_____

b) Analyze a **counterargument** to YOUR position and defend the reasons why somebody might believe other arguments besides your own.

_____
_____
_____
_____

2) Clarify how **investigative spectroscopy** used by astronomer Vesto Slipher (1875-1969) was able to determine that **distant galaxies** are **redshifted**, and by doing so provided the first empirical evidence for the **expansion of the universe** in 1917.

_____
_____
_____
_____
_____

3) Differentiate between the viewpoints on the universe by Albert Einstein (1879-1955) and Georges Lemaître (1894-1966).
   a) Einstein – **static universe**

_____
_____
_____

   b) Lemaître - **Big Bang theory, expanding universe, primeval atom (cosmic egg)**

_____
_____
_____

4) Explain the reasons why **distant galaxies** appear to **recede** away from us.

_____
_____
_____
_____

5) Clarify how **the Big Bang** model hypothesis about the formation of the universe got its name from astronomer Fred Hoyle (1915-2001).

_____
_____
_____
_____
_____
_____

6) Analyze the existence of the **lookback time** that not only explains the farther away something is the farther in the past we see it, but also the **redshift** of **distant galaxies** receding into the **background glow** of the **fireball** left over from **the Big Bang**.

_____
_____
_____
_____
_____
_____
_____

7) Analyze the reasons why **the Big Bang** model of the universe has essentially been proven to be correct. _Include how **the Big Bang** model includes the relationship of **hydrogen** to **helium**._

_____
_____
_____
_____
_____
_____

8) Evaluate how **the Big Bang** was the initial exploding and **expansion of space** itself by explaining how **space in the universe expanded rapidly** rather than the idea of **matter** rushing away from **some point in space**.

_____
_____
_____
_____
_____
_____

9) Determine how it is possible for scientists and astronomers to know with a high degree of certainty that the **age of the universe** is around **13.82 billion years old**.

_____
_____
_____
_____
_____

Name_____
Period_____
Date_____

Dark Energy, Cosmology part 2: Crash Course Astronomy #43

1) Explain the idea surrounding the existence of an **infinitely dense point** that came about before **the Big Bang** called **the Singularity**.

_____
_____
_____
_____

2) Clarify how **mutual gravity** is causing the **Andromeda Galaxy (M31)** to speed towards the **Milky Way Galaxy** at about 50 km/sec instead of moving away from us as a result of our expanding **universe**.

_____
_____
_____
_____
_____

3) Evaluate the reasons why **white dwarf type 1A supernovae** make good **standard candles** in the **distant galaxies** that they inhabit. *Include how comparing this information to **supernovae redshifts** lets astronomers calculate not only their **distance** but also how fast the **universe** is expanding.*

_____
_____
_____
_____
_____

4) Connect the reasons why astronomers came to the conclusion that the **universe** was **accelerating** rather than slowing down as expected. *Include how this is due to the **gravity** from all the **matter** in the **universe**.*

_____
_____
_____
_____

5) Clarify how 95% of the **universe** is **matter** that cannot be seen on account of **dark energy expansion** which in turn is accelerating the **expansion of the universe**.

_____
_____
_____
_____

6) Hypothesize as to whether or not the **geometry of the universe** will cause the **expansion of the universe** to continue forever. *Defend your answer by analyzing hypothetical scenarios.*

_____
_____
_____
_____

7) Defend the hypothesis that there may be enough **dark energy** with **negative pressure** (repulsive action) in **space** to ensure that the **expansion of the universe** will continue forever. *Include what YOU think this would mean for the **fate of the universe**.*

_____
_____
_____
_____

8) Connect the reasons why at some point in the future, **distant galaxies** could be moving **faster than the speed of light** away from our location in space as space itself is exempt from the **cosmic speed limit**.

_____
_____
_____
_____

9) Clarify the reasons why astronomers can currently see **galaxies** about 45 billion light years away in the radius of the **observable universe** even though the **cosmological expansion** of the universe is only 13.8 billion years old.

_____
_____
_____
_____

10) Explain how the **light of a galaxy** is fighting more and more expansion all the time due to the inflationary forces of **dark energy**.

_____
_____
_____
_____

11) ***Bonus Question***
Hypothesize how measuring the **brightness** of **red giants** in **distant galaxies** could resolve the debate over the rate of **expansion of the universe**. *Include the differences between the **rate of comoving distance** when using **Cepheid variables** to using numbers obtained from the expansion of the space in the **Cosmic Microwave Background**.*

_____
_____
_____
_____

A Brief History of the Universe: Crash Course Astronomy #44

1) Explain how experiments in **giant particle colliders** have provided astronomers with a lot of information about the **early universe**.

_____

_____

_____

2) Clarify why the **early universe** was unfathomably hot and dense at its start. *Include how molecules were affected along with their **atoms** and even their **nuclei**.*

_____

_____

_____

3) Clarify how **subatomic particles** called **quarks** are formed. *Include if **quarks** and electrons are **basic particles** or if they can be subdivided even further.*

_____

_____

_____

_____

4) Analyze how the **cosmos** was a bizarre and unfamiliar place that pretty much consisted of an unbelievably **hot soup** of **electrons** and **quarks** in the beginning of the universe.

_____

_____

_____

_____

5) Analyze what happened to the **four forces of physics** (i.e. **gravity, electromagnetism, weak force, strong force**) after the first second that the universe existed.

_____

_____

_____

6) Analyze how **atoms** formed three to seventeen minutes after that the universe existed.

_____

_____

_____

_____

7) Briefly explain why **fusion** stopped leaving the primordial ratio of **3:1 hydrogen to helium atoms** 20 minutes after that the universe existed.

_____
_____
_____
_____
_____

8) Explain how the intense heat of the early **opaque universe** caused the **ionization** of **atoms** all the way up to **370,000 years** after the **Big Bang**.

_____
_____
_____
_____
_____
_____

9) Analyze how the **recombination event** of the **transparent universe** which occurred about 370,000 years after the **Big Bang** is responsible for the **light** we see as the **cosmic microwave background** today.

_____
_____
_____
_____
_____
_____

10) Clarify how astronomers know that **matter** was very evenly distributed everywhere as well as having the same temperature in the universe on account of the **smooth light** emitted by the **cosmic microwave background** 13.8 billion years ago.

_____
_____
_____
_____
_____
_____

11) Evaluate the proposal of **inflation** that theoretical physicist Alan Guth (1947-   ) added to supplement the **Big Bang** model which included a massive increase of **velocity**.

_____
_____
_____
_____
_____
_____

Deep Time: Crash Course Astronomy #45

1) Clarify how using **scientific notation** can be useful when using large numbers especially when one is talking about **dates and time** in the **far future** of the universe.

_____
_____
_____
_____
_____

2) Briefly analyze the **five broad epochs** as defined in the book *The Five Ages of the Universe* (1999) by Fred Adams (1961-   ) and Greg Laughlin (1967-   ):

a) **Primordial Era - Big Bang - primordial soup, nucleosynthesis, recombination of electrons**

_____
_____
_____
_____
_____
_____

b) **Stelliferous Era** – 10^{13} years – current stars until only **red dwarfs, galaxies** fade

_____
_____
_____
_____
_____
_____

c) **Degenerate Era** – 10^{40} years - dark – **infrared** only **neutron stars, white dwarfs, brown dwarfs, binary brown dwarfs, proton decay**

_____
_____
_____
_____
_____

d) **Black Hole Era** - 10^{92} years – **black hole** evaporation, **black holes** leaking **mass**

_____
_____
_____
_____
_____

3) Analyze what will happen after the **Black Hole Era** or at the **end of the universe** aptly named the **Dark Era**. *Defend the reasons as to why the Dark Era could stretch on to infinity*.

_____
_____
_____
_____
_____

4) Analyze the plausible idea of the existence of **dark energy** which MIGHT prevent this potential eternal darkness inside of the **Dark Era** from happening to the **universe**.

_____
_____
_____
_____
_____
_____
_____

5) Evaluate the feasibility of the **Big Rip** that MIGHT happen as a result of **dark energy** if the **cosmic horizon** shrinks until it is smaller than a **subatomic particle**.

_____
_____
_____
_____
_____
_____

6) Critique the possibility that we live in a **multiverse**, and that our **universe** is just one of many **infinite universes**. *Defend your answer by hypothesizing whether or not other universes may survive even though our universe has wound down*.

_____
_____
_____
_____
_____
_____

7) Hypothesize a scenario in which there could be a **cosmic reboot** where the **laws of physics** could be rewritten to create an entirely **new universe**.

_____
_____
_____
_____
_____
_____

Everything, the Universe...And Life: Crash Course Astronomy #46

1) Explain why finding **oxygen** on **exoplanets** that are close to the size of **Earth** and located in the **habitable zone** around its **host star** would be a critical observation and key to finding **extraterrestrial life**.

_____

_____

_____

_____

2) Describe the types of **exoplanets** that have been found so far in the 21st century.

_____

_____

_____

_____

3) Evaluate the reasons why most astronomers in the 21st century believe that most **stars** in the **galaxy** have **planets**, whereas most astronomers in the 20th century thought that the planets in the **solar system** were unique.

_____

_____

_____

_____

_____

4) Analyze how **liquid water** covering the surface of **Earth** early in its history strongly implies that getting **life** started on planets is easy.

_____

_____

_____

_____

5) Differentiate between **extraterrestrial life** and **intelligent extraterrestrial life**.
   a) **Extraterrestrial life** – simple lifeforms like **bacteria** or **algae**

   _____

   _____

   _____

   b) **Intelligent extraterrestrial life** (complex life – plants, fungi, fish, animals, or **self-aware beings** with **intelligent thought**)

   _____

   _____

   _____

   _____

6) Briefly evaluate the problems with **interstellar travel** in a **spaceship** between **stars** and their **planets**.

_____
_____
_____
_____
_____

7) Analyze the premise behind **SETI (Search for Extraterrestrial Intelligence)**, and why **SETI** believes that **radio waves** are the best way to communicate with **aliens**. _Include how chatting with **aliens** on the **radio** may be easier than flying in a **spaceship** to meet them._

_____
_____
_____
_____
_____

8) Hypothesize as to how **intelligent aliens** (if they exist) might **communicate** with each other in ways which humans may not be able to understand. _Think of the ways bees, plants, or fungi communicate with each other as a basis for YOUR hypothesis._

_____
_____
_____
_____

9) Predict whether or not YOU think that **humans** will make **contact** with either **intelligent life** or **microbial life** within the next 20 years or even within YOUR lifetime. _Defend your answer with logical assumptions._

_____
_____
_____
_____

10) Briefly evaluate how questions raised in **astronomy** can provoke a sense of wonder and awe. _Include anything in **astronomy** that has provoked a sense of wonder and awe to YOU and evaluate the reasons why._

_____
_____
_____
_____
_____
_____

ESSAYS

1) Lunar Colonization
2) The Case to Colonize Mars
3) Cloud City on Venus
4) Asteroid Mining: Costly Endeavor or Profitable Adventure?
5) Mining Mercury to Create a Dyson Sphere
6) Space Stations Made of Space Junk
7) The Fermi Paradox and Abiogenesis: Finding Alien Life, a Good Thing?
8) Utilizing LaGrange Points Between Earth and the Moon
9) Planet X: Trans-Neptunian Object or Primordial Black Hole?
10) Spaceship Designs

1) **Lunar Colonization** - Predict how **lunar colonization** will happen on the **Moon** by analyzing the **three phases of colonization** which have occurred on **Earth** in the past. *Use the **colonizing of the Americas** as the basis for your **hypothesis** and then relate the **colonization of the Americas** to **lunar colonization**:
   a) **Expeditions** to the **Moon**
   b) **Outposts** and **Settlement**s founded
   c) True **Colony Self-Sufficiency** & **Economically Productive Lunar Society** (e.g. exporting **helium 3**, **hydrogen-rich lunar water** for **habitats** and **rocket fuel**)

_____
_____
_____
_____
_____
_____
_____
_____
_____
_____
_____
_____
_____
_____
_____
_____
_____
_____
_____
_____
_____
_____
_____
_____
_____
_____
_____
_____
_____
_____
_____
_____
_____

Lunar Colonization

_____
_____
_____
_____
_____
_____
_____
_____
_____
_____
_____
_____
_____
_____
_____
_____
_____
_____
_____
_____
_____
_____
_____
_____
_____
_____
_____
_____
_____
_____
_____
_____
_____
_____
_____
_____
_____
_____
_____

2) The Case to Colonize Mars

Make a case for going to and living on **Mars**.

*Synthesize a scenario that includes how your proposition will be able to overcome the many life-threatening dangers on the planet using plausible **science** and **technology**.*

- *Example **Pros**: Solid surface, **oxygen** extraction for **life support** from local **carbon dioxide** possible, possible or probable ancient or **current bacterial-type life**; Space tourism, **interplanetary species**, mining possibilities, **metal-rich** moons **Phobos** and **Deimos**, advanced **Martian geographical knowledge**, possible **base location** to launch missions to the **asteroid belt** and outer planets.*

- *Example **Cons**: Oxygen for **life support** needs to be converted from **carbon dioxide**, **nitrogen poor**, dangerous levels of **radiation** on surface of **Mars**, cold planet, accessing **water**, **growing food** may be difficult, very **thin atmosphere** comparable to 1% of **Earth atmosphere**, **pressurized habitats** and **space suits** needed, only **38% Earth-like gravity**, **radioactive dust**, no **magnetic field**, **psychological effects** of **physical containment** and the effects of **isolation in space**.*

_____
_____
_____
_____
_____
_____
_____
_____
_____
_____
_____
_____
_____
_____
_____
_____
_____
_____
_____
_____
_____
_____
_____
_____
_____
_____
_____
_____

The Case to Colonize Mars

_____
_____
_____
_____
_____
_____
_____
_____
_____
_____
_____
_____
_____
_____
_____
_____
_____
_____
_____
_____
_____
_____
_____
_____
_____
_____
_____
_____
_____
_____
_____
_____
_____
_____
_____
_____
_____
_____
_____

3) Cloud City on Venus

*Analyze the pros and the cons to creating **floating habitable structures** and **cities** on **Venus**. Defend your answer with possible **scientific, technological,** and **economic endeavors** that might take place on the planet.*

- *Example **Pros**: **Earth-like pressure** (at sea level), **temperature,** and **radiation intake** in **sulfur-laced clouds 50-60 km above surface of Venus**. **Earth-like gravity,** possible **terraforming** adventure. **Hydrogen, Oxygen,** and **Nitrogen** would "float" on top of the **clouds of Venus** making it easier for **balloons** to float on **Venus** than **Earth**. **Nitrogen-rich** and **Carbon Dioxide-rich** for trade in the future **solar economy**.*

- *Example **Cons**: **No habitable surface** (without **terraforming**), outside **oxygen levels insufficient** to breathe, **sulfuric acid** present in clouds. **Crushing pressure** (93 bars) **and intense heat** (740K) **on surface** due to **runaway greenhouse effect**.*

_____

_____

_____

_____

_____

_____

_____

_____

_____

_____

_____

_____

_____

_____

_____

_____

_____

_____

_____

_____

_____

_____

_____

_____

_____

_____

_____

_____

_____

_____

Cloud City on Venus

Asteroid Mining: Costly Endeavor or Profitable Adventure in the **Solar Economy**.
*Analyze the **cost efficiency** of mining **near-Earth asteroids (NEOS)** in the foreseeable future. Include why a company or nation might want to **mine an asteroid**.*

_____
_____
_____
_____
_____
_____
_____
_____
_____
_____
_____
_____
_____
_____
_____
_____
_____
_____
_____
_____
_____
_____
_____
_____
_____
_____
_____
_____
_____
_____
_____
_____
_____
_____
_____
_____
_____
_____

Asteroid Mining

_____
_____
_____
_____
_____
_____
_____
_____
_____
_____
_____
_____
_____
_____
_____
_____
_____
_____
_____
_____
_____
_____
_____
_____
_____
_____
_____
_____
_____
_____
_____
_____
_____
_____
_____
_____

4) **Mining Mercury to Create a Dyson Sphere**. *Hypothesize a scenario in which it might be possible to harness the power **of the Sun** by building an **energy-capturing structure** around it.*

_____
_____
_____
_____
_____
_____
_____
_____
_____
_____
_____
_____
_____
_____
_____
_____
_____
_____
_____
_____
_____
_____
_____
_____
_____
_____
_____
_____
_____
_____
_____
_____
_____
_____
_____
_____
_____

Mining Mercury to Create a Dyson Sphere

_____

_____

_____

_____

_____

_____

_____

_____

_____

_____

_____

_____

_____

_____

_____

_____

_____

_____

_____

_____

_____

_____

_____

_____

_____

_____

_____

_____

_____

_____

_____

_____

_____

_____

_____

_____

_____

_____

_____

5) **Building Space Stations Out of Scrapped Rockets and Space Junk**: Foolish idea or ingenious plan. *Evaluate the benefits and drawbacks of a potential plan to get rid of all the **junk orbiting Earth in space** by using it to make **habitats** and **space stations**.*

_____
_____
_____
_____
_____
_____
_____
_____
_____
_____
_____
_____
_____
_____
_____
_____
_____
_____
_____
_____
_____
_____
_____
_____
_____
_____
_____
_____
_____
_____
_____
_____
_____
_____
_____
_____

Building Space Stations Out of Scrapped Rockets and Space Junk

_____
_____
_____
_____
_____
_____
_____
_____
_____
_____
_____
_____
_____
_____
_____
_____
_____
_____
_____
_____
_____
_____
_____
_____
_____
_____
_____
_____
_____
_____
_____
_____
_____
_____
_____
_____
_____

6) **The Fermi Paradox and Abiogenesis: Finding Alien Life: A Good Thing?** *Critique the problems with finding other examples of **abiogenesis** in the universe such as microbial, fish-like, or intelligent life on other planets or moons using the logic of **The Fermi Paradox**.*

_____
_____
_____
_____
_____
_____
_____
_____
_____
_____
_____
_____
_____
_____
_____
_____
_____
_____
_____
_____
_____
_____
_____
_____
_____
_____
_____
_____
_____
_____
_____
_____
_____
_____
_____
_____

The Fermi Paradox and Abiogenesis: Finding Alien Life: A Good Thing?

_____
_____
_____
_____
_____
_____
_____
_____
_____
_____
_____
_____
_____
_____
_____
_____
_____
_____
_____
_____
_____
_____
_____
_____
_____
_____
_____
_____
_____
_____
_____
_____
_____
_____
_____
_____
_____
_____
_____
_____

8) **Utilizing LaGrange Points Between Earth and the Moon**

   A) Evaluate how utilizing the five **Lagrange Points** located around the Earth and the Moon will play a strong role in colonizing **cislunar space**.

   B) Analyze how the creation of **infrastructure** for the **exploration of space** would work at some or all of these points due to the **gravitational balance** created there by fluctuations of space between the **Earth** and the **Moon**. *Include why those locations will be important points to park **weigh stations, habitats, factories, relay satellites,** and **space wheels** for those interested in **colonizing space.***

   i) **Lagrange Point #1 – halo orbits**, ideal for **personnel** and **cargo station**, unstable, **orbital station keeping**, easy **transfer** into **lunar** or **Earth orbit**

   ii) **Lagrange Point #2 – halo orbits**, unstable, ideal for **relay stations**

   iii) **Lagrange Point #3** – not easy to use

   iv) **Lagrange Point #4** – stable, convenient points for a **habitat**

   v) **Lagrange Point #5** – stable, convenient points for a **habitat**

   _____
   _____
   _____
   _____
   _____
   _____
   _____
   _____
   _____
   _____
   _____
   _____
   _____
   _____
   _____
   _____
   _____
   _____
   _____
   _____
   _____
   _____
   _____
   _____
   _____
   _____
   _____
   _____

Utilizing LaGrange Points Between Earth and the Moon

_____
_____
_____
_____
_____
_____
_____
_____
_____
_____
_____
_____
_____
_____
_____
_____
_____
_____
_____
_____
_____
_____
_____
_____
_____
_____
_____
_____
_____
_____
_____
_____
_____
_____
_____
_____
_____
_____
_____
_____

9)**Planet X: Trans-Neptunian Object** or **Primordial Black Hole**
Evaluate the reasons why gravitational anomalies inside the **Kuiper Belt** near some **dwarf planets** and other icy objects indicate the possible existence of a **Neptune-sized object** or a **primordial black hole**. *Hypothesize as to how it is quite possible that there is a **9th planet** or a **primordial black hole** beyond **Neptune** due to the orbits of **some trans-Neptunian objects** and **long-period comets** which have very similar **perihelia**.*

_____
_____
_____
_____
_____
_____
_____
_____
_____
_____
_____
_____
_____
_____
_____
_____
_____
_____
_____
_____
_____
_____
_____
_____
_____
_____
_____
_____
_____
_____
_____
_____
_____
_____
_____

Planet X: Trans-Neptunian Object or **Primordial Black Hole**

_____
_____
_____
_____
_____
_____
_____
_____
_____
_____
_____
_____
_____
_____
_____
_____
_____
_____
_____
_____
_____
_____
_____
_____
_____
_____
_____
_____
_____
_____
_____
_____
_____
_____
_____
_____
_____

10)Spaceship Designs

Evaluate the effectiveness of different **spaceship designs** that would be successful at both **interplanetary** and **interstellar travel**. *Be sure to consider and evaluate the factors of travelling through the vacuum of space such as **radiation shielding, spaceship shape, habitat modules, gravity, freighters** without humans, **A.I. (Artificial Intelligence)**, and types of **fuel** that would be needed to power each type of spaceship.*

_____
_____
_____
_____
_____
_____
_____
_____
_____
_____
_____
_____
_____
_____
_____
_____
_____
_____
_____
_____
_____
_____
_____
_____
_____
_____
_____
_____
_____
_____
_____
_____
_____
_____

Spaceship Designs

_____
_____
_____
_____
_____
_____
_____
_____
_____
_____
_____
_____
_____
_____
_____
_____
_____
_____
_____
_____
_____
_____
_____
_____
_____
_____
_____
_____
_____
_____
_____
_____
_____
_____
_____
_____
_____
_____

*More Crash Course Books written by Roger Morante and currently for sale on Amazon include:

1) Crash Course Psychology: A Study Guide of Worksheets for Psychology

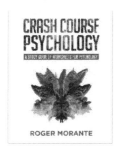

2) Crash Course Biology: A Study Guide of Worksheets for Biology

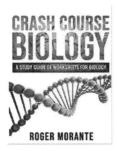

3) Crash Course Anatomy: A Study Guide of Worksheets for Anatomy

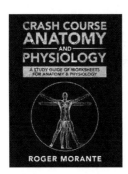

4) Crash Course US History: A Study Guide of Worksheets for US History

5) Crash Course Government and Politics: A Study Guide of Worksheets for Government and Politics

6) Crash Course World History: A Study Guide of Worksheets for World History

7) Crash Course Economics: A Study Guide of Worksheets for Economics

8) Crash Course Literature: A Study Guide of Worksheets for Literature

9) Crash Course Philosophy: A Study Guide of Worksheets for Philosophy

10)Crash Course World History II: A Study Guide of Worksheets for World History II

11)Crash Course European History: A Study Guide of Worksheets for European History

*Please email morante13@crashcourse.org.in to contact the publisher.
*Access to answer keys available upon proof of purchase. Must be 18 years or older to access.
*Visit my public Facebook page and follow me for the most up to date publications at:

Roger Morante
@rogermorante13

Made in the USA
Columbia, SC
17 August 2022

65549011R00070